# The Reference Shelf®

U.S. National Debate Topic: 2025–2026

# Exploration & Development of the Arctic

The Reference Shelf
Volume 97 • Number 3
H.W. Wilson
a Division of EBSCO Information Services, Inc.

Published by
**GREY HOUSE PUBLISHING**
Amenia, New York
2025

# The Reference Shelf

Cover photo: iStock.

The books in this series contain reprints of articles, excerpts from books, addresses on current issues, and studies of social trends in the United States and other countries. There are six separately bound numbers in each volume, all of which are usually published in the same calendar year. Numbers one through five are each devoted to a single subject, providing background information and discussion from various points of view and concluding with an index and comprehensive bibliography that lists books, pamphlets, and articles on the subject. The final number of each volume is a collection of recent speeches. Books in the series may be purchased individually or on subscription.

Copyright © 2025 by Grey House Publishing, Inc. All rights reserved. No part of this work may be used or reproduced in any manner whatsoever or transmitted in any form or by any means, electronic or mechanical, including photocopying, recording, or any information storage and retrieval system, without written permission from the copyright owner. For subscription information and permissions requests, contact Grey House Publishing, 4919 Route 22, PO Box 56, Amenia, NY 12501.

∞ The paper used in these volumes conforms to the American National Standard for Permanence of Paper for Printed Library Materials, Z39.48 1992 (R2009).

Publisher's Cataloging-in-Publication Data
(Prepared by Parlew Associates, LLC)

Names: Grey House Publishing, Inc., compiler.
Title: U.S. national debate topic : 2025-2026 : exploration & development of the Arctic / [compiled by Grey House Publishing].
Other Titles: Exploration & development of the Arctic. | Exploration and development of the Arctic. | Reference shelf ; v. 97, no. 3.
Description: Amenia, NY : Grey House Publishing, 2025. | Includes bibliographic references and index. | Includes color and b&w photos and illustrations.
Identifiers: ISBN 9798891793194 (v. 97, no. 3) | ISBN 9798891793163 (volume set)
Subjects: LCSH: Arctic regions – Climate. | Arctic regions – Environmental conditions. | Arctic regions - Economic conditions. | Sustainable development – Arctic regions. | Geopolitics – Arctic regions. | Arctic peoples. | BISAC: POLITICAL SCIENCE / Geopolitics. | SCIENCE / Global Warming & Climate Change. | POLITICAL SCIENCE / Indigenous / General.
Classification: LCC G606.G74 2025 | DDC 338.998—dc23

Printed in Canada

# Contents

Preface ix

# 1

## Introduction to the Arctic

Understanding the Arctic 3

Explainer: The Geopolitical Significance of Greenland 10
Jennifer Spence and Elizabeth Hanlon, *Belfer Center for Science and International Affairs*, January 16, 2025

Changing Geopolitics in the Arctic 15
Esther D. Brimmer
*Council on Foreign Relations*, July 18, 2023

No. 26 | NATO in the Arctic: 75 Years of Security, Cooperation, and Adaptation 20
Elley Donnelly, *Wilson Center*, April 3, 2024

Fulfilling the Promise of the Arctic Council 23
Paul Fuhs, *Northern Forum*, 2025

The Golden Age of Offbeat Arctic Research 30
Paul Bierman, *Undark*, September 6, 2024

# 2

## Indigenous Interests

Independent Action in the Arctic 37

Many People in the Arctic Are Staying Put Despite Climate Change, Study Reports 43
Katie Bohn, *Penn State University Agricultural Sciences*, May 10, 2024

The True Cost of Mining in the Canadian Arctic 46
Henry Harrison, *The Circle (WWF)*, April 2023

Valuing Indigenous Knowledge in Permafrost Research 49
Meral Jamal, *Undark*, January 10, 2024

We Went to Greenland to Ask about a Trump Takeover 55
Ben Schreckinger, *Politico*, January 10, 2025

No. 21 | The Arctic Council and the Crucial Partnership Between
  Indigenous Peoples and States in the Arctic                               65
Edward Alexander and Evan T. Bloom, *Wilson Center*, July 27, 2023

Understanding the Arctic Through Indigenous "Perspectives"                  71
Marta Asenjo Fernández, *REVOLVE*, July 9, 2024

# 3

## Sustainable Economic Development

The Search for Arctic Resources                                             79

There's a Global Tug-of-War for Greenland's Resources—But the
  New Government Has Its Own Plans                                          86
Nicolas Jouan, *The Conversation*, March 24, 2025

How Trump's Looming Tariffs May Impact Canada's Arctic                      89
Hilde-Gunn Bye, *High North News*, February 5, 2025

Critical Minerals in the Arctic: Forging the Path Forward                   92
Brett Watson, Steven Masterman, and Erin Whitney, *Wilson Center*, July 10, 2023

Unleashing Alaska's Extraordinary Resource Potential                        99
Donald Trump, *The White House*, January 20, 2025

The Arctic Trilemma: The United States Must Compete in the
  Transpolar Sea Route                                                      105
Ashton Basak, *Georgetown University Center for Security Studies*, October 5, 2024

# 4

## Climate and Environmental Protection

The Arctic Ecosystem                                                        115

We Need Greenland. But Not in the Way Trump Thinks                          122
Paul Bierman, *Undark*, January 23, 2025

Arctic Has Changed Dramatically in Just a Couple of Decades—2024
  Report Card Shows Worrying Trends of Snow, Ice, Wildfire, and More        126
Twila A. Moon, Matthew L. Druckenmiller, and Rick Thomas, *The Conversation*,
  December 10, 2024

Climate Collaborations in the Arctic Are Frozen Amid War                    134
Jessica McKenzie, *Undark*, April 5, 2022

Climate Change Is Fuelling Trump's Desire to Tap into Canada's
    Water and Arctic Resources      138
Tricia Stadnyk, *The Conversation*, January 20, 2025

The Oil Industry's Cynical Gamble on Arctic Drilling      141
Rebecca Leber, *Vox*, September 8, 2023

Greenland Is Getting Greener—Helped by a Mining Company and a
    Group of Tree Enthusiasts      145
Adriana Craciun, *The Conversation*, December 6, 2024

# 5

## Arctic Militarization

The Strategic North      157

Rising Tensions and Shifting Strategies: The Evolving Dynamics of
    US Grand Strategy in the Arctic      163
Kiel Pechko, *The Arctic Institute*, January 7, 2025

India's Arctic Challenge: Aligning Strategic Interests with Regional
    Realities      173
Nima Khorrami, *The Diplomat*, September 30, 2024

Resource Wars: How Climate Change Is Fueling Militarization of
    the Arctic      176
Joanna Rozpedowski, *RealClear Defense*, August 7, 2024

Why the US Is Losing the Race for the Arctic and What to Do About It      180
Josh Caldon, *Center for International Maritime Security (CIMSEC)*, April 13, 2023

Bibliography      185
Websites      189
Index      191

# Preface

## The Arctic Circle
The Arctic, a term derived from the Greek word for "bear," comprises the northernmost part of the Earth. This unique polar region, characterized by extreme weather, ice, and cold, has shaped life in unique ways. From the polar bears, seals, and whales that evolved over millions of years to thrive in this unique frozen world, to the many Indigenous cultures that, over thousands of years, developed cultures and societies based on their relationship to the ice and to the Arctic Sea.[1]

Long a source of fascination for the people who came from regions further south around the world, exploration of the Arctic ultimately revealed both a fantastic wealth of resources and strategic territory with global geopolitical significance. Gradually, the nations with Arctic territories, including Canada, Norway, Russia, Sweden, Finland, and Iceland, came to colonize and occupy the Arctic, wresting control from the Indigenous inhabitants of these territories. Denmark and the United States also came to control Arctic territory through their acquisition of Greenland and Alaska, respectively. In the years that followed, resource exploitation and development transformed the Arctic, while the industrialization of the larger world changed the environment of Earth as a whole.[2]

By the twenty-first century, the Arctic had been transformed. Warming temperatures eliminated more than half of the ice that once covered the ocean and landmasses, oil and gas prospecting polluted the air and drove animals to extinction. In the midst of this, resource gathering companies found new opportunities, and enhanced access to potential wealth and resources. This brough the Arctic back into relevance as an increasingly important frontier for resource harvesting and national military defense strategies. With these emerging concerns stirring debate, politicians, scientists, and activists are debating how best to manage the hunger for resources and global power against the welfare of this vast but rapidly diminishing landscape.[3]

## Arctic Origins
The story of the Arctic begins with the migrants who were the first to come to the Arctic from Asia. Waves of migration created a variety of different cultures, some of which still exist in the twenty-first century. There are more than forty different Indigenous ethnic groups that still live in the Arctic, comprising 10 percent of the overall population. These include well-known groups like the Sáami, famous for their herding of reindeer, and also includes the Aleut, Inuit, and Yupik people, all once called "eskimos" by Europeans outside of the Arctic. European habitation of

the Arctic began with the Viking mariners, like Eric the Red, who discovered the island that is now known as "Greenland" around 980. In the 1600s, European explorers like William Baffin, Henry Hudson, and the legendary Captain Cook, discovered new locations in the Arctic, and with them made first contact between Europe's monarchies and the millennia-old Indigenous societies in this part of the world.[4]

Trade from the Arctic began with fish and sea mammals, and this also led to the first trade disputes as participating nations soon worried that overexploitation would lead to shortages in the supply of valuable resources, like whale oil, which once powered lamps across Europe. In the 1800s, the world shifted to the petroleum economy, and Arctic oil was discovered. This changed the trajectory for the Arctic and the Arctic people, both Indigenous and colonial alike. From the nineteenth century to the modern era, mining has been the chief global aim in the Arctic, with Canada, Russia, and the other Arctic nations investing heavily in both the extraction of fossil fuels and other mineral resources. This effort has accelerated as global warming has reduced the sea ice and diminished the glaciers, making it easier for corporations to access these subterranean resources.

The militarization of the Arctic can be traced back to the efforts to protect the seafood trade, but accelerated during the World Wars, when the Arctic became a strategic route for supplies and a battleground. In the Cold War that followed, the Arctic was a staging ground for the buildup of nuclear weapons, and this contest continues in the twenty-first century, though discussion of nuclear armaments is no longer as prominent in the media or in public discourse as it once was.[5]

## The Modern Arctic

Once a realm in which sea ice and glaciation made navigation difficult or in some cases impossible, the Arctic is changing and with these changes companies and military administrators are perceiving new opportunities and potentially new military threats. This had led to renewed discussion about the Arctic region and the potential benefits and risks of Arctic resource exploration. First, the Arctic has returned to public discourse is the disappearance of Arctic sea ice and glaciers due to climate change. Second, the status of resource supplies in other parts of the world has driven interest in capitalizing on newly available Arctic resources.

Management of the Arctic depends on the governments of the countries that possess Arctic territories. Russia is the nation with the most Arctic territory, with the Russian coastline comprising more than 50 percent of the Arctic landmass as a whole. Russia also has the largest population of residents living in any of the Arctic territories, and thus Russia has the highest level of overall interest in the management of the Arctic.[6] Further, the Russian economy is highly dependent on the fossil fuel economy, and this makes the Arctic region among the most important territorial advantages for the Russian state.

The Scandinavian nations collectively control significant Arctic territories, which have been important to the survival of Indigenous Arctic communities and at the forefront of the effort to control climate change and protect Arctic ecosystems. One third of the landmass of Sweden is within the Arctic circle, though it remains one of the least-populated regions in the nation. Sweden's Sáami people live in the northern portion of the Arctic and have long played an active role in managing the region's resources. Nearly half of the land controlled by the nation of Norway is within the Arctic circle, and around one-tenth of the Norwegian population live in the Arctic region. Like Sweden, Norway's Arctic territories are the traditional home of the Sáami people of the region, whose population also spreads into Finland and Russia. A small portion of Iceland, Grimsey Island, located 40 kilometers off the Icelandic coast, falls within the Arctic circle. Iceland has been aggressive in environmental regulation in its portion of the Arctic, efforts that are unable to stem the tide of global changes in the environment. Around one-third of Finland is above the Arctic Circle, including the entire province of Lapland. Finland's Arctic reaches are sparsely populated, but is a refuge for the nation's Sáami population.

In North America, both Canada and the United States have claim over some portion of Arctic territory. Canada's claim is far more robust as nearly 40 percent of the landmass of Canada is within the Arctic. However, though Canada has vast Arctic territory, less than 1 percent of the Canadian population is settled in this region. More than half of the 150,000 Canadians who live in the Arctic are Indigenous people. The United States acquired Arctic interests with the 1867 purchase of Alaska and, as such, have only a disconnected connection to the Arctic region. The state of Alaska is the largest in the United States, but one of the least populated. This sparse population is at odds with the commercial importance of the Alaskan Arctic, which has been heavily focused on petroleum production.

Finally, the Kingdom of Denmark has a disconnected interest in the Arctic through the semi-independent nation of Greenland, which was once a colonial territory of Denmark. Denmark continues to be closely involved in the administration of Greenland in terms of foreign relations and economics, making Greenland residents citizens of the European Union (EU), but the independence movement in Greenland has been pushing for further autonomy. Greenland has become a major focus of modern debates on the Arctic after Donald Trump suggested that the United States should seek to acquire Greenland as a new state and/or territory. The vast majority of Greenland residents are against this proposal and there are no international legal standards that support this proposal becoming reality without major long-term shifts in international dynamics. Trump's claims have been perceived as an act of aggression and this has filtered into the broader debate.

All of the nations that have permanent territory in the Arctic have participated in the Arctic Council, an intergovernmental organization that works to research the Arctic in various ways and to manage relations between participating countries. The 2022 Russian invasion of the Ukraine forced the Arctic Council to alter operations, suspending Russian leadership and participation. The strength and function of the council has been further thrown into question by the election of Donald Trump, whose attitude towards the organization is not well understood, but whose general posturing with regard to the Arctic conflicts with many of the policies and positions held by other member nations.

## The Arctic Debate

The debate over the Arctic and management of the Arctic has many dimensions, but wealth and power are at the root of the issue. The Trump administration's suggestion of acquiring Greenland, and the Russian military's recent buildup of arms and naval operations in the Arctic, are examples of the great powers posturing and demonstrating interest in claiming additional Arctic influence as access to Arctic resources increase. This increase in interest then raises major questions about the function and facility of existing international agreements, territories, boundaries, and trade. For the United States, Canada, and Russia, and to a lesser extend the Scandinavian nations, access to petroleum resources is paramount and thus shapes the political aspects of this debate in many different ways. In the United States, this means corporate influence on behalf of the petroleum mining and processing companies that stand to increase profit with additional resources.

On the other side of this debate are environmentalists trying to preserve Arctic ecosystems and wildlife. Activists and scientists have been studying the Arctic and its environment in an effort to better understand global climate change and also to illustrate the importance of Arctic ice as a buffering force in controlling the global climate. Environmental activism overlaps with Indigenous activism and organizations, who are trying to ensure the welfare of Indigenous Arctic residents, in part by preserving ecosystems, local lifestyles, and an influential voice in the global debate over the region's potential resources. The Arctic debate is likewise shaped by many nongovernmental organizations (NGOs), some representing environmental interests, others studying and providing policy guidance on the economic future of the region.

## Works Used

"The Arctic." *Arctic Centre, University of Lapland*, 2025, www.arcticcentre.org/EN.

"Arctic Countries." *Arctic Review*, 2025, arctic.review/international-affairs/arctic-countries/#:~:text=their%20Arctic%20policies.-,Russia,important%20stake%20in%20the%20region.

"Arctic Exploration Timeline." *American Polar Society*, 2024, americanpolar.org/arctic-exploration-timeline/.

Dunaway, Finis. *Defending the Arctic Refuge: A Photographer, an Indigenous Nation, and a Fight for Environmental Justice*. U of North Carolina P, 2021.

Laineman, Matti, and Juha Nurminen. *A History of Arctic Exploration: Discovery, Adventure and Endurance at the Top of the World*. Bloomsbury USA, 2009.

Murkins, Sydney. "The Future Battlefield is Melting: An Argument for Why the U.S. Must Adopt a More Proactive Arctic Strategy." *Arctic Institute*, 3 Dec.

2024, www.thearcticinstitute.org/future-battlefield-melting-argument-us-must-adopt-more-proactive-arctic-strategy/.

Reid, Robert Leonard. *Arctic Circle: Birth and Rebirth in the Land of the Caribou.* David Godine Publishers, 2010.

## Notes

1. Reid, *Arctic Circle.*
2. "The Arctic," *Arctic Centre, University of Lapland.*
3. Dunaway, *Defending the Arctic Refuge.*
4. Laineman and Nurminen, *A History of Arctic Exploration.*
5. Murkins, "The Future Battlefield Is Melting."
6. "Arctic Countries," *Arctic Review.*

# 1
# Introduction to the Arctic

Arctic Council Ministerial Meeting, Reykjavík, Iceland, 2021. Photo by U.S. Department of State, via Wikimedia. [Public domain.]

# Understanding the Arctic

The Arctic is a vast polar region at the northernmost end of the Earth marked by cold and extreme habitats and unique animal and plant communities adapted to live in this challenging environment. Many modern nations have territories that extend into the Arctic region, including Norway, Sweden, Finland, Russia, Canada, the United States, and Iceland. The Arctic region is connected to the Arctic Ocean, the smallest of the world's oceans, and the coldest, and ultimately part of the Atlantic Oceanic system. The North Pole is found in the Arctic, as is the Bering Strait, a key location in the history of humanity. Inhospitable and difficult for long-term human habitation, the Arctic has long been a refuge of natural resources, though the oceans and landscape of the Arctic have also been repeatedly exploited for resources.

Humans discovered the vast polar region known as "the Arctic" at least 14,000 years ago, when travelers from Eastern Asia traveled over the Bering Land Bridge, which connected Eastern Asia to Canada, and from there to what is now the United States. Another wave of migration occurred around 5,000 years ago, with the arrival of a group known to archaeologists as the "Paleo-Eskimos." Modern Indigenous people of Siberia, Alaska, and the Aleutian Islands can trace their ancestry back to this wave of Paleo-Eskimos. One recent study, looking at ancestry in the Arctic, found that one of the first populations in the Arctic territory of Greenland came there from Siberia around 3000 BCE and were isolated in that territory for more than 4,000 years, ultimately disappearing from the archaeological record.[1]

Archaeological evidence indicates that most of the Paleo-Eskimo populations consisted of small villages each containing perhaps twenty to thirty people. Bill Fitzhugh, of the National Museum of Natural History, explained that the longevity of these societies suggests a very traditional lifestyle closely linked to the land. "One might almost say, kind of jokingly and very informally, that [they] were the hobbits of the Eastern Arctic." These arrivals and later arrivals from Asia, including the "Thule" people, most likely came by boat, and the land bridge that brought the first visitors to the Arctic was gone by this time.[2]

Scientists aren't certain how long this land bridge was navigable, but Earth was entering one of many warming periods that have occurred in geological history. The sea levels rose and the land bridge disappeared between 11,000 and 13,000 years ago. It is important to understand that while periods of warming—where glaciers melt and sea levels rise—are a natural part of Earth's history. The current warming trend, called "global warming" or "climate change," is not part of this cycle. The current warming of the climate is being caused by the mismanagement of natural resources and the burning of fossil fuels, and it is happening far faster than any natural warming trend. The result, ultimately, will be a devastating loss of species, habitats, and resources. The situation is especially dire in the Arctic, where the loss of glacial ice is transforming ecosystems in ways that negatively impact animals and humans living in the region.

In any case, a combination of land migrations through the Bering Land Bridge, and oceanic migration in sea vessels of some kind, resulted in the population of the Arctic over thousands of years and for further thousands of years, Indigenous Arctic societies thrived and survived, remaining closely tethered to the unique natural environments of this region.

## Creating the European Arctic

In the 330s, BCE, a Greek explorer and geographer named Pytheas came across the Arctic while traveling north from Britain. He was the first European known to have described this landscape, though records of his exploration survive only indirectly, through the writings of others who read them. The next recorded encounter with the Arctic didn't come until 1497, when Italian explorer John Cabot found parts of what is now North America when searching for a better route through to Asia.

Most famous among the early explorers who visited this snowy realm is Martin Frobisher, who set out from England and undertook three separate expeditions to the Arctic between 1576 and 1578. It was Frobisher who whet European appetites for the Arctic region when he reported the discovery of gold on Baffin Island, now part of the Canadian territory of Nunavit. Frobisher called the island "Queen Elizabeth's Foreland," and there is a bay on the island, "Frobisher Bay," named for the explorer himself. The European term for the island was established in 1616 in honor of another English explorer, William Baffin, who rediscovered the island while searching for a sea lane connecting the Atlantic and the Pacific. The name for the island among the Indigenous Inuktitut people who live there is Qikitaaluk.[3]

The promise of gold on Baffin Island, or Qikitaaluk drew explorers from Europe, and though the gold that Frobisher though he found ended up being pyrite, European settlement of the Arctic begins with Frobisher, who gave England their first territorial claims on what became the nation of Canada.

Though Europe now had a foothold, there really wasn't much to be done in the Arctic before oil was discovered and the region was too isolated for harvesting resources. However, in the mid-1800s, Arctic exploration intensified and was driven by scientific curiosity. Among the most famous voyages was the expedition led by Sir John Franklin in 1845, still seeking the potential passage from the Atlantic to the Pacific. Franklin, along with 128 men on the HMS *Erebus* and the HMS *Terror*, disappeared. A reward was offered, leading to a search-and-rescue effort that stretched for years. It wasn't until 2014 that the wreck of the *Erebus* was discovered, with the wreck of the *Terror* found a couple of years later.

This was the event that first brought American explorers to the Arctic and though they didn't discover the fate of Franklin's crew (and so didn't claim the reward) this began the industry of American whale hunting, harvesting whale oil and fat to be shipped back to the Americas. This was why the United States became interested in the Alaskan territory and ultimately purchased Alaska from Russia in 1867 for a cost of $7.2 million, without any consultation with or treaties established with the native inhabitants of the Arctic territory.

The famed Northwest Passage was finally discovered and successfully navigated, by Finnish scientist Adolf Erik Nordenskiöld in 1878, and this ended the race for the Northwest Passage, but also began another international contest, with explorers from North America and Europe racing to be the first to reach the North Pole. This led to another event, in 1909, when explorers Robert Peary and Matthew Henson claimed to have reached the North Pole. Their claim came just after Frederick Cook, another American explorer, made the same claim. Though the claim was doubted and disputed the event stands out in history if for no other reason than because Matthew Henson, an African American explorer, was one of the first nonwhite explorers to make a major name for himself in Europe, and the first to visit the Arctic. It is unclear if any of these men actually reached the pole and it wasn't until 1968, when American explorer Ralph Plaisted reached the geographic pole by snowmobile that this feat could be confirmed.

While science drove the early 1900s explorations of the Arctic, both industry and science drove interest in the Arctic moving forward. Expeditions had confirmed that the Arctic was rich in minerals, and gold was discovered in Alaska's Yukon range in the 1890s. The 1900s was a period of resource exploration, alongside scientific studies in an effort to understand the glaciers, the movement of oceanic and wind currents, the formation of the continents, and many other mysteries that had intrigued humans for thousands of years. Ultimately, Russia discovered coal, diamonds, nickel, and copper in the Arctic territories that Russia controlled, while the United States and Russia discovered oil and gas reserves. Gas production began in Alaska's Prudhoe Bay in 1968, and in Siberia in the 1970s. Since then, Canada and Norway have likewise tapped into these ancient resources, fueling their economies while also contributing the pattern of climate change that now poses an existential threat to the world.

## Militarization of the Arctic

World War II made the Arctic a strategic military location. Arctic military bases were used to store weapons and other military supplies, to establish weather stations, and airstrips, and there were battles in the Arctic as the Germans attacked some of the Allied bases located in the region. When the end of the war came, the Arctic was a changed place. Littered with military bases, the Arctic became a staging ground and a focal point in the Cold War that followed. There were radar and surveillance stations, military facilities of many different kinds and, ultimately, nuclear weapons. Because much of the Arctic is sparsely populated, Arctic military outposts became a frequent site for nuclear weapons tests. From 1955 to 1990, Russia detonated eighty-eight atmospheric, twenty-nine underground, and three underwater nuclear devices.

Early in the nuclear arms race the developed nations of the world established what is called "mutually assured destruction," or MAD. What this means, essentially, is that none of these nations can actually use a nuclear weapon on another nation without being destroyed in the process. Nuclear weapons are therefore a threat, rather than a useable weapon. Despite this, military advocates have continued to push for continued nuclear weapons development. This is often a strategy of distraction, where citizens are lulled into believing that further nuclear technology will provide them with some enhanced safety or advantage so that citizens, reacting from fear, allow their government to invest huge sums of money into nuclear technology even as this investment does little to nothing to enhance safety or security. Russia, a militant authoritarian regime, continues to develop nuclear facilities in the Arctic, despite gaining no appreciable military advantage for doing so. This military buildup is claimed to be necessary to protect economic interests in the region. Russia does not gain military strength from this development, but some analysts have suggested that Russia may intend to essentially force foreign cooperation as other nations take an interest in preventing future nuclear hazards and waste buildup in the region.[4]

## Access and Cooperation in the Arctic

Prior to the Second World War, the Arctic Ocean was divided according to the "sector theory," which was introduced in Canada in 1907. The basic idea was simple, straight lines drawn from the Arctic countries to the North Pole, and this would serve as the fair division of the Arctic. The Soviet government adopted the Sector theory in 1926, while the United States remained largely disinterested in disagreements that primarily impacted the Soviet Union and Europe.[5]

The threat of global human annihilation became very real during the Cold War in the 1980s, but this ultimately created incentive, for both the United States and Russia (then the Soviet Union) to come to some agreements on *sharing* at least part of the Arctic territories. In the 1990s, the super powers negotiated the United Nations Convention on the Law of the Sea (UNCLOS), an international governing body that could help to solve territorial disputed with regard to the Arctic. The United States was the only country that never ratified the treaty, though key aspects of the treaty were incorporated into US tradition, if not US law. According to this system, a state can claim a 12-nautical-mile territory sea, and a 200-nautical-mile exclusive economic zone, extending from the naturally occurring borders of the landmass in question. To make these determinations, it is necessary to precisely measure the length and extent of the continental shelf surrounding each landmass in the Arctic, and this is what scientists have done over the ensuing years, helping to establish claims to territory in international law.

The 1990s also saw the establishment of the Arctic Environmental Protection Strategy (AEPS), which was signed by the eight countries involved; the United

States, Canada, Denmark, Finland, Iceland, Russia, Norway, and Sweden. In 1996, with the Ottawa Declaration, these signatory nations created the Arctic Council, an intergovernmental body that would allow participating countries to discuss shared issues and opportunities in the Arctic region. Indigenous communities were likewise given a voice and some degree of influence in the Arctic Council as well as scientific organizations. Over the years that have followed, the Arctic Council has produced studies on fossil fuel deposits in the Arctic Region and has been involved in the negotiation of treaties regarding shipping and transport. The first binding treaty accepted by member states did not come until 2011, a system for coordinating search and rescue in the case of emergencies in the Arctic.[6]

After the Russian invasion of the Ukraine, in 2022, the Arctic Council essentially suspended activities as the remaining member nations refused to work with Russia, then leading the council, on any further projects. Later that year, the other nations of the Arctic Council resumed work on projects without the inclusion of Russia. Russia's exclusion, and the reelection of Donald Trump has left the Arctic Council's future uncertain and tenuous.

During Trump's first term in office, the Arctic Council, for the first time in the organization's history, failed to agree on a final declaration on climate change, due to the fact that then Secretary of State Mike Pompeo refused to accept the term "climate change." The Trump administration has embraced climate change denial as a core policy position, eliminating spending and efforts to combat this process. Another major focus of the Arctic Council has been to enhance and support Indigenous voices and influence in the Arctic. Given the Trump administration's hostility towards efforts to support diversity, equity, and inclusion, it seems unlikely that the administration will demonstrate any concern for the Indigenous welfare function of the Arctic Council either. The Trump administration is also friendly with Russia, despite the Russian government's continued violations of international law in Ukraine.

In addition, the Trump administration has repeatedly put forward the idea that the United States might "purchase" or otherwise come to control Greenland, a semiautonomous Arctic country that was once part of Denmark. These coinciding factors have led many analysts with knowledge of the area to suggest that the United States will either derail or, at least, will not contribute positively to the cooperative administration of the Arctic Council moving forward.[7]

## The Future of the Arctic

More than at any time in the past, the future of the Arctic is uncertain, both politically and environmentally. The lives and future welfare of the Indigenous residents of the Arctic States also stand on shaky ground. While there are many issues, including the rise in militarism and economic uncertainty, there is no more pressing issue than climate change, which threatens the economies, lifestyles, and natural resources of the Arctic, but may simultaneously ease the barriers of access to Arctic fossil fuels, and this is the issue upon which the entire global debate over the Arctic ultimately hinges.

The world's largest and most powerful nations have proven unwilling to seriously consider combating climate change, because doing so would diminish profit in the petroleum industry and, ultimately, the wealthy class that dominates the political environment in these countries. Corporations attempting to preserve profit fund inaccurate "studies" and reports that cast doubt on scientific data, and support the careers of politicians who encourage doubt in scientific consensus and/or promote the idea that preservation of current corporate sources of revenue is more directly connected to the quality of working-class life than changes in the climate or equity in human society. Climate change is an overarching source of concern, in part, because it shapes all the other issues occurring in the Arctic region, and also because the preservation of the Arctic glaciers are tantamount to humanity's ability to address this problem before facing the most severe after-effects.

## Works Used

"Ancient DNA Sheds Light on Arctic Hunger-Gatherer Migration to North American around 5,000 Years Ago." *Science Daily*, 5 June 2019, www.sciencedaily.com/releases/2019/06/190605133522.htm.

Griggs, Mary Beth. "The First People to Settle Across North America's Arctic Regions Were Isolated for 4,000 Years." *Smithsonian Magazine*, 28 Aug. 2014.

Heidt, Amanda. "Did Humans Cross the Bering Strait after the Land Bridge Disappeared?" *Live Science*, 30 Dec. 2023, www.livescience.com/archaeology/did-humans-cross-the-bering-strait-after-the-land-bridge-disappeared#:~:text=According%20to%20John%20Hoffecker%2C%20a,during%20one%20of%20these%20stretches.

Huebert, Rob. "Can the Arctic Council Survive the Trump Administration? Probably Not. Here's Why." *Arctic Today*, 3 Mar. 2025, www.arctictoday.com/can-the-arctic-council-survive-the-trump-administration-probably-not-heres-why/.

Jóhannesson, Magnús. "Arctic Council: Structure, Work and Achievements." *Arctic Circle*, 12 Dec. 2022, www.arcticcircle.org/journal/arctic-council-structure-work-and-achievements.

"Qikiqtaaluk Region." *Government of Nunavut*, 2024, www.gov.nu.ca/sites/default/files/documents/2024-08/03_-_Info_about_Qikiqtaaluk_Region.pdf.

Sharp, Gregor. "A Brief History of Lines in the Arctic." *Arctic Institute*, 20 Mar. 2018, www.thearcticinstitute.org/brief-history-lines-arctic/.

Starchak, Maxim. "Russia's Arctic Policy Poses a Growing Nuclear Threat." *Carnegie Endowment*, 1 Nov. 2024, carnegieendowment.org/russia-eurasia/politika/2024/10/russia-arctic-nuclear-threat?lang=en.

## Notes

1. Heidt, "Did Humans Cross the Bering Strait after the Land Bridge Disappeared?"
2. Griggs, "The First People to Settle Across North America's Arctic Regions Were Isolated for 4,000 Years."
3. "Qikitaaluk Region," *Government of Nunavut*.
4. Starchak, "Russia's Arctic Policy Poses a Growing Nuclear Threat."
5. Sharp, "A Brief History of Lines in the Arctic."
6. Jóhannesson, "Arctic Council."
7. Huebert, "Can the Arctic Council Survive the Trump Administration?"

# Explainer: The Geopolitical Significance of Greenland

By Jennifer Spence and Elizabeth Hanlon
*Belfer Center for Science and International Affairs*, January 16, 2025

## Introduction

President Donald Trump has repeatedly reiterated his interest in acquiring Greenland, either through economic coercion or military force, describing it as vital to safeguarding U.S. national security and countering the growing influence of China and Russia in the Arctic. U.S. House Republicans are rallying around this idea, and in a recent visit to Greenland, U.S. Vice President JD Vance accused Denmark of not doing a good job keeping Greenland safe.

Greenland's Prime Minister Múte B. Egede rebuked Trump's comments, stating that "Greenland is for the Greenlandic people," with Danish Prime Minister Mette Frederiksen echoing his sentiment.

Trump's design on Greenland isn't new: he proposed buying the island during his first term, drawing widespread ridicule. But what exactly explains Trump's continued interest in Greenland?

## Who Governs Greenland?

### Self-Governing

Greenland, the world's largest island with a majority-Inuit population of about 56,000, is a self-governing country within the Kingdom of Denmark. Greenland's government manages most domestic matters, including education, health, and natural resource development. While the Danish government has final say over foreign, defense, and security policy, Greenland's autonomy in these areas is growing. In 2024, Greenland released its Foreign, Defense, and Security Strategy 2024–2033, titled *Greenland in the World—Nothing About Us Without Us*.

### Moving Toward Independence

After previously being a colony and then a province of Denmark, Greenland gained self-rule in 1979. The 2009 Greenland Self-Government Act expanded Greenland's responsibilities and gave Greenlanders the right to declare independence from Denmark. Most Greenlanders support eventual independence, though economic reliance on Danish subsidies complicates this goal.

---

From *Belfer Center for Science and International Affairs*, January 16 © 2025. Reprinted with permission. All rights reserved.

## Why Is Greenland Considered a Strategic Location?

### Arctic Geopolitics

Since the end of the Cold War, the Arctic has generally been an area of international cooperation; however, climate change, resource competition, and growing militarization, especially by Russia, have raised geopolitical tensions in the region over the last decade. Russia's invasion of Ukraine in 2022 splintered its relations with the other seven Arctic states (Canada, Kingdom of Denmark, Finland, Iceland, Norway, Sweden, and the United States) and prompted Finland and Sweden to join NATO in 2023 and 2024 respectively. As a result, all Arctic states except Russia are NATO members. This shift has elevated the overall importance of the Arctic—including Greenland, which is by default part of the alliance through the Kingdom of Denmark—to NATO.

> "With the right to self-determination and the goal of independence, our country and people aim to increase their cooperation with other countries. It is important for us as responsible citizens of the world, in our own name, to have the courage to take a stand on issues and events around the world."
> —*Greenland in the World–Nothing About Us Without Us*

### U.S. Military Capabilities

Greenland hosts Pituffik Space Base, formerly Thule Air Base, a U.S. military installation key to missile early warning and defense as well as space surveillance. Greenland is also part of the GIUK Gap (Greenland-Iceland-United Kingdom), an anti-submarine warfare chokepoint in the North Atlantic during the Cold War, which today remains important for monitoring and potentially restricting Russian naval movements in the North Atlantic and Arctic Ocean. However, the island's strategic military value to the United States has waned since the end of the Cold War due to evolutions in military technology, and investment in Pituffik Space Base has been sporadic over the years.

### Future Trans-Arctic Shipping

Greenland occupies a key position along two potential shipping routes through the Arctic: the Northwest Passage, along the northern coastline of North America, and the Transpolar Sea Route, through the center of the Arctic Ocean. As Arctic sea ice melts, these routes could reduce shipping times and bypass traditional chokepoints like the Suez and Panama Canals. Currently, these routes are commercially unviable and will likely remain so for many years because of treacherous weather and floating ice. In the long term, as vessel traffic in the Arctic Ocean increases, Greenland will likely become a key player in effective management of the Arctic Ocean, including emergency management, prevention, and response. The viability of these new shipping routes and other maritime activities in the region will depend, among other things, on investments in comprehensive marine infrastructure. Greenland is strategically positioned to both benefit from and help manage such investments.

## Why Are Other Countries Interested in Greenland's Natural Resources?

Melting land and sea ice is making Greenland's rich mineral and hydrocarbon deposits more accessible, attracting the notice of countries competing for energy transition materials. However, developing these resources remains challenging due to Greenland's harsh environment, remoteness, lack of infrastructure, high extraction costs, and local concerns about environmental impacts and risks to traditional livelihoods.

### Critical Minerals

Greenland is a potential source of many minerals that are critical for the energy transition. In particular, Greenland has large deposits of rare earth elements (REEs) required for manufacturing batteries, wind and solar technologies, and advanced military equipment. Western countries see Greenland's mineral resources as an opportunity to reduce their dependence on China, which dominates critical mineral supply chains. China has also expressed interest in Greenland's mineral wealth, but a Chinese-backed REE project stalled after Greenland banned uranium mining. Moreover, serious doubts around the economic feasibility of mineral resource extraction persist due to Greenland's harsh environment, remoteness, lack of infrastructure, and high operating costs. As of 2023, Greenland had just two active mines, with a handful of other projects in development. Mining is also controversial among Greenlanders: while some see mining as a path toward independence, others worry about the impacts on the environment and traditional subsistence livelihoods. However, they are united in their position that Greenlanders will decide.

**Greenland occupies a key position along two potential shipping routes through the Arctic: the Northwest Passage, along the northern coastline of North America, and the Transpolar Sea Route, through the center of the Arctic Ocean.**

### Oil and Gas

According to a 2007 U.S. Geological Survey report, there could be significant oil and gas reserves off Greenland's coast. However, Greenland's government stopped issuing new licenses for oil and gas exploration in 2021, citing concerns about the economic feasibility and environmental impacts of drilling.

### Fresh Water

Commercial exploitation of Greenland's ice and water could help meet growing global demand for fresh water. Approximately twenty percent of the world's fresh water is locked up in the ice sheet that covers the island.

### Fisheries

Greenland's fisheries are crucial for its economy, providing livelihoods for local communities, being Greenland's main export commodity, and contributing significantly to the nation's GDP. Greenland is an important contributor to the global supply of fish and, with climate change, its significance may actually increase.

## Is Greenland Influenced by China?

### Failed Foothold

China has shown interest in Greenland's mineral wealth and proximity to potential shipping routes, but in recent years its presence on the island has dwindled. In 2018, China released a white paper detailing its Arctic strategy, including its intention to build a "Polar Silk Road," in parallel with its Belt and Road Initiative infrastructure investments in other regions. During the 2010s, Greenland courted Chinese mining companies to invest, but subsequent mining projects involving Chinese partners have stalled or failed. Pressure from the United States also helped quash Chinese bids to construct new airports and convert an abandoned Danish naval base into a research station. Though Greenland has expressed openness to working with international partners, China has not renewed its overtures. China's main presence in the Arctic is in Russia. Greenland's largest trading partner is China, but this fact is not necessarily significant by itself, since China is the largest trading partner for most countries in the world.

## What Is Getting Lost in the Current Conversation about U.S. Interests in Greenland?

### Self-Determination

The Greenland Self-Government Act, ratified by the Danish Parliament in 2009, recognizes the Greenlandic people as "a people pursuant to international law with the right to self-determination." Greenland cannot be acquired by the United States—or any other country—without the consent of Greenlanders.

### Climate and Environment

The melting Greenland ice sheet is one of the largest contributors to global sea level rise, making the island a focal point for understanding the global climate system. In addition, Greenland contains globally unique ecosystems and biodiversity hotspots, such as the North Water Polynya, an area of year-round open water and one of the most biologically productive areas in the Arctic Ocean. Since Indigenous and local knowledge is crucial for understanding changes in Greenland's climate and environment and related impacts, countries should partner with Greenland's well-established research community to support high-quality, locally based research. Greenland published its *National Research Strategy*

> "Greenlandic researchers and institutions are already doing high-quality work, and we need to continue to harness the potential of our human, technological and logistical resources to enhance our capabilities in research and sustainable development."
>
> —Peter Olsen, Greenland's *National Research Strategy 2022–2030*

2022–2023 and is developing ethical guidelines for international researchers seeking to conduct research in Greenland.

**Reality Versus Potential**

Opportunities for mining and trans-Arctic shipping will not be commercially viable in the near term.

## Conclusion

It is for Greenlanders to decide the future of Greenland. The United States does not need to own Greenland to achieve its security and economic interests, and Trump's current approach may be counterproductive to those interests. His threats may have the unintended consequences of alienating Greenland, the Kingdom of Denmark, and other U.S. allies and undermining both a long tradition of cooperative Arctic governance and the existing norms of the current international system.

(*Note:* This piece has been updated following President Trump's inauguration and his reaffirmation of ambitions around Greenland.)

> "Is he serious? Would he do it? These are the questions being floated in the media. Somehow we have rapidly moved from how inappropriate this proposition is to examining its feasibility and likelihood. . . . The United States will not be made safer by dominating its neighbors. Security in the Arctic will not be achieved through acts of aggression against U.S. allies. Global stability will not be sustained if the rules-based order becomes optional."
>
> —Jennifer Spence, *Trump Renews Pursuit of Greenland*

### Print Citations

**CMS**: Spence, Jennifer, and Elizabeth Hanlon. "Explainer: The Geopolitical Significance of Greenland." In *The Reference Shelf: U.S. National Debate Topic 2025–2026: Exploration & Development in the Arctic,* edited by Micah L. Issitt, 10–14. Amenia, NY: Grey House Publishing, 2025.

**MLA**: Spence, Jennifer, and Elizabeth Hanlon. "Explainer: The Geopolitical Significance of Greenland." *The Reference Shelf: U.S. National Debate Topic 2025–2026: Exploration & Development in the Arctic,* edited by Micah L. Issitt, Grey House Publishing, 2025, pp. 10–14.

**APA**: Spence, J., & Hanlon, E. (2025). Explainer: The geopolitical significance of Greenland. In M. L. Issitt (Ed.), *The reference shelf: U.S. national debate topic 2025–2026: Exploration & development in the Arctic* (pp. 10–14). Grey House Publishing. (Original work published 2025)

# Changing Geopolitics in the Arctic

By Esther D. Brimmer
*Council on Foreign Relations*, July 18, 2023

Before the Subcommittee on Transportation and Maritime Security United States House of Representatives 1st Session, 118th Congress: Hearing on "Strategic Competition in the Arctic"

Thank you, Chairman Gimenez, Ranking Member Thanedar, and members of the subcommittee, for inviting me to testify today about "Strategic Competition in the Arctic."

The Arctic sits at the confluence of three phenomena: shifting geopolitics, changing climate, and the far-ranging implications of Russia's invasion of Ukraine. The Arctic is geographical, the home of almost four million people facing the impact of climate change that will alter lives and livelihoods. It is also conceptual. Just as the words "Indo-Pacific" or the "South China Sea" connote strategic concepts, so too the "Arctic" takes on renewed strategic meaning. The Arctic is "America's Fourth Coast" meriting increased attention to the interlocking strategic, economic, environmental, and social concerns in this region.

The Arctic Circle begins at 66.5°N (north of the equator). Eight countries have territory in the Arctic Circle: Canada, Finland, Denmark, Iceland, Norway, the Russian Federation, Sweden, and the United States. The end of the Cold War reduced political pressures in the Arctic region. The spirit of the Norwegian concept, "High North, Low Tension" prevailed. The strategic situation has mutated into a new configuration. By 2023, renewed great power competition around the world is manifest in the Arctic region.

The Arctic intensifies the effects of decisions made elsewhere. The geopolitics of the Arctic were dramatically altered by the Russian Federation's invasion of Ukraine in 2022. Russia's invasion of Ukraine fundamentally transformed the security calculations of two longtime neutral countries. As a result of Russia's actions, Nordic states Finland and Sweden applied for membership in the North Atlantic Treaty Organization (NATO). Finland had been neutral since 1948 and Sweden had been neutral since the Napoleonic era two centuries ago. Both abandoned neutrality to seek the security of the world's most powerful military alliance.

This expansion recalibrates politics within NATO. With the accession of Finland, six (and with Sweden seven) of the Arctic countries are formal allies. Finland was admitted in April 2023, bringing NATO an 832-mile land border with

---

From *Council on Foreign Relations*, July 18 © 2023. Reprinted with permission. All rights reserved.

Russia. The upshot for the Arctic is that the region transmutes from a region with five NATO allies, two strategically neutral states, and the Russian Federation to a zone with potentially seven NATO allies and the Russian Federation.

NATO is a defensive military alliance, but it is also a framework for deep cooperation among the national security communities of the member states. Henceforth, the Arctic will play a larger role in the strategic operations, calculations, and exercises of America's most important military alliance. The North Atlantic and Arctic would be important for North American supplies flowing to European allies in a crisis. The institutions of the alliance will increasingly embed Arctic and High North topics into their work. For example, Allied Command Transformation states, "...the High North is an important priority for NATO" when explaining the addition of Arctic activities to its projects preparing NATO members for future challenges. Presidential time is valuable. The importance of the High North was exemplified by President Joe Biden's trip to Helsinki for the United States-Nordic Leaders' Summit after the July 11-13, 2023, NATO summit. Furthermore, Finland and Sweden are both members of the European Union, making two more EU members also NATO members, which could alter EU security discussions.

Russia's invasion of Ukraine not only enhanced NATO, it also inadvertently stalled cooperation in one of the Arctic's most distinctive multilateral organizations: the Arctic Council. Founded in 1996 in the afterglow following the end of the Cold War, the Arctic Council embodies the spirit of cooperation; decisions are made by consensus. The forum focuses on "sustainable development and environmental protection in the Arctic." By design, the Arctic Council does not address security issues. The Council has adopted three legally-binding agreements: the Agreement on Cooperation on Aeronautical and Maritime Search and Rescue in the Arctic (2011), the Agreement on Cooperation on Marine Oil Pollution Preparedness and Response in the Arctic (2013), and the Agreement on Enhancing International Arctic Scientific Cooperation (2017).

Unusual for an intergovernmental body, the Arctic Council also includes six Permanent Participants representing Arctic Indigenous Peoples. This special facility for interaction is distinctive and should be preserved. Cultural ties span current national borders. Indigenous peoples have lived in the harsh climate of the Arctic for over a thousand years; their expertise and perspectives can be relevant as countries seek to understand climate change.

At the time of the 2022 invasion of Ukraine, Russia happened to hold the rotating chairmanship of the Arctic Council. As part of the international response to the invasion, the other seven members of the Arctic Council paused cooperation with Russia in that body. Upon assuming the two-year chairmanship in May 2023, Norway sought to revitalize cooperation in the Arctic Council articulating four priorities: "the oceans; climate and environment; sustainable economic development; and people in the north." Another venue for cooperation, the Arctic Coast Guard Forum remains dormant with Russia holding the chairmanship through 2023.

This strategic realignment in the Arctic builds on political shifts that were already evident before the invasion. Recent years witnessed a resurgence of great power competition. The United States faces a rising power, China, and the Russian Federation. Increasingly, countries outside the Arctic have become more active in the region. China called itself a "near-Arctic" state in its 2018 Arctic Policy White Paper. In 2013, China, Japan, India, Italy, the Republic of Korea, and Singapore became Arctic Council Observers, joining France, Germany, The Netherlands, Poland, Spain, Switzerland, and the United Kingdom.

Many countries and companies are interested in access to resources. The Arctic is home to living and mineral resources. Managing access in the fragile Arctic environment is challenging. Yet, agreements are possible. Arctic countries share a concern about illegal, unreported, and unregulated fishing that depletes delicate natural resources and vulnerable wildlife. Canada, China, the Kingdom of Denmark (in respect of the Faroe Islands and Greenland), Iceland, Japan, the Republic of Korea, Norway, the Russian Federation, the United States, and the European Union are parties to the Agreement to Prevent Unregulated High Seas Fisheries in the central Arctic Ocean, which entered into force in 2021 and initially will be in force until 2037. The agreement would be automatically extended for another five years as long as none of the Parties object.

In 2008, the U.S. Geological Survey estimated that 13 percent, or 90 billion barrels, of the world's undiscovered conventional oil resources were in the Arctic. Most of these resources are in Alaska and the Russian Federation. The Arctic plays an important role in the Russian economy. About half of the Arctic area is Russian coastline. Twenty percent of Russia's land mass is in the Arctic Circle and includes large cities. Russia wants others to use (and pay to use) the Northern Sea Route.

Even before the war in Ukraine, Russia needed partners for economic development. Economic sanctions promulgated as part of the international response to Russia's invasion of Ukraine foreclose options for Russia.

Russia's need for investment opens a gateway for China to be more involved in Arctic issues. *High North News* notes that China has invested $90 billion in energy and resource projects in the Arctic over the past decade, largely in Russia. China is Russia's leading trade partner, as China is for 120 countries. China's investments in the Arctic are related to its Belt and Road Initiative. Yet, patterns of Chinese shipping were different in 2022. *High North News* reports that whereas China's COSCO shipping company had been the largest non-Russian operator along the Northern Sea Route (NSR), it did not send any ships along the NSR in 2022. In 2022, of the 314 ships sailing along the Northern Sea Route, only thirty-six were non-Russian-flagged vessels. Nevertheless, Chinese investment in Russia continues to grow. Chinese-Russian trade rose to a "record $190 billion" in 2022. There are European countries that still have economic links with Russia. European Union countries' consumption of Russian LNG increased 50 percent since sanctions started, mostly going to Belgium, France, and Spain.

Increased activity by China and Russia in the Arctic is a manifestation of another trend: great power competition in global spaces. For over a century the United States has enjoyed command of the seas and more recently airspace and outer space. Access to sea routes, airwaves, cyberspace, and satellite information are all necessary for modern economies to function, but also require using shared international spaces that may be beyond or at the edges of national jurisdiction. In many parts of the world great power and assertive middle powers seek access to resources, some of which may be in or under these global spaces. Access to the global commons and areas beyond national jurisdiction is crucial for success in an era of strategic and commercial rivalry. Therefore, protection of coastlines, waterways, safe commercial transit, and management of marine resources place extra demands on the United States Coast Guard.

Oceans are especially sensitive. At the center of the Arctic region is the Arctic Ocean, which is beyond the jurisdiction of any country. The United Nations Convention on the Law of the Sea creates the international legal regime for oceans, including the Arctic Ocean. Each Arctic country, including the U.S., claims its 200-mile exclusive economic zone. The U.S. is at a disadvantage because it is not a party to the United Nations Convention on the Law of the Sea, which provides mechanisms for countries to claim more rights. Canada, Russia, and Denmark (on behalf of Greenland) turned to one of those mechanisms, the United Nations Commission on the Limit of the Continental Shelf (CLCS) regarding their overlapping claims to the Lomonosov Ridge under the Arctic Ocean. The CLCS made non-binding recommendations in February 2023 about the extent of Russia's claim. Further diplomatic or legal work will need to occur to settle the borders.

**Climate change is important to the geopolitics of the Arctic because it changes access to the oceans. There could be ice-free summers in the Arctic Ocean in the 2030s. Companies and countries watch to see if navigation through the Arctic would be viable.**

The Arctic, like other regions of the world, benefits from layers of global governance. Even in an era of geopolitical upheaval, cooperation on technical standards facilitates commercial, social, and environmental interactions. The International Maritime Organization's International Code for Ships Operating in Polar Waters (Polar Code), which entered into force in 2017, provides important standards for shippers operating in the Arctic and Antarctic regions. The terms of the Polar Code are mandatory under both the International Convention for the Safety of Life at Sea (SOLAS) and the International Convention for the Prevention of Pollution from Ships (MARPOL).

Fundamental to understanding the geopolitical and economic issues in the Arctic is the phenomenon of climate change. Global warming is occurring in the Arctic possibly three times as fast as in the rest of the world. Sea ice is frozen seawater. With less Arctic sea ice to reflect sunshine away from the Earth, the planet will continue to heat up. Furthermore, the Greenland ice sheet (which is frozen freshwater) has lost ice for the past twenty-five years.

The ongoing geopolitical shifts occurring before the invasion of Ukraine were premised on climate change. Climate change is important to the geopolitics of the Arctic because it changes access to the oceans. The warming climate means that more areas of the Arctic are ice-free in the summer, possibly opening opportunities for navigation. There could be ice-free summers in the Arctic Ocean in the 2030s. Companies and countries watch to see if navigation through the Arctic would be viable, thereby shortening shipping routes and times between Asia and Europe. Other observers counter that even with less ice, Arctic navigation would still be difficult.

Climate change challenges livelihoods. Around 4 million people live in the Arctic, and about 2 million of them are Russian; about 500,000 are Indigenous people. Around sixty percent of Alaska Native communities are "environmentally threatened" by climate change. Conditions are especially acute for Indigenous people who still hunt for sustenance. Thin ice and altered animal migrations mean hunters must travel farther for food. Migration patterns of birds and fish, and also caribou, walruses, and whales have shifted, requiring people to extend the hunting season. Warmer waters may entice fish usually found in lower latitudes to move farther north. The changing climate also affects companies' calculations. Shell ended offshore exploration in Alaska in 2015.

The Biden Administration's October 2022 National Strategy for the Arctic Region includes investments in the Arctic. To advance maritime security in an era of strategic competition in the Arctic, the United States must continue to deepen its commitment to

- Make progress on building a deep-water port in Nome, Alaska.
- Continue the Polar Security Cutter program.
- Work with the current chair of the Arctic Council, Norway, to sustain mechanisms that promote human and environmental well-being, including connections among Indigenous Peoples in the Arctic region.

**Print Citations**

**CMS**: Brimmer, Esther D. "Changing Geopolitics in the Arctic." In *The Reference Shelf: U.S. National Debate Topic 2025–2026: Exploration & Development in the Arctic*, edited by Micah L. Issitt, 15–19. Amenia, NY: Grey House Publishing, 2025.

**MLA**: Brimmer, Esther D. "Changing Geopolitics in the Arctic." *The Reference Shelf: U.S. National Debate Topic 2025–2026: Exploration & Development in the Arctic*, edited by Micah L. Issitt, Grey House Publishing, 2025, pp. 15–19.

**APA**: Brimmer, E. (2025). Changing geopolitics in the Arctic. In M. L. Issitt (Ed.), *The reference shelf: U.S. national debate topic 2025–2026: Exploration & development in the Arctic* (pp. 15–19). Grey House Publishing. (Original work published 2023)

# No. 26 | NATO in the Arctic: 75 Years of Security, Cooperation, and Adaptation

By Elley Donnelly
*Wilson Center*, April 3, 2024

As NATO commemorates its 75th birthday on April 4th, we celebrate the alliance's steadfast commitment to a peaceful and prosperous Arctic. The High North has been intricately woven into NATO's trajectory since its inception, embodying a nexus of geopolitical importance and strategic interests that have profoundly influenced the alliance's evolution.

Formed in 1949 in collective response to the threat of Soviet aggression, NATO's foundational membership included Arctic states such as Canada, Norway, Denmark (Greenland), Iceland, and the United States. These countries held national interests in the Arctic, and understood the strategic significance of the region, its potential as a military theater, and its role in global maritime trade. The participation of these states laid the groundwork for NATO's consideration of the Arctic as a crucial frontier for security interests from the outset.

The Arctic's unique geography is characterized by unforgiving icy waters, remote islands, and vast expanses, which have continually influenced NATO's strategic calculus. During World War II, the Arctic theater saw the tactical importance of weather control, known as the Weather War, where Allied and Axis powers sought to manipulate weather patterns to gain military advantage in the region. The proximity of NATO's Arctic boundaries to Russia's northern borders further underscores the region's significance as a potential flashpoint for geopolitical competition.

## Cold War Era: Tension in the Arctic

During the Cold War, the Arctic emerged as a critical frontline in the confrontation between NATO and the Soviet Union. The distinctive Arctic landscape, including the strategic Greenland-Iceland-UK (GIUK) Gap, proved conducive to intelligence gathering and asset positioning. The Arctic Ocean also served as a potential route for Soviet submarines and bombers, heightening NATO's concerns. The increased military activity and perceived threat prompted NATO to establish a robust defensive presence in the region. The Cold War era witnessed the deployment of early warning systems, military bases, and surveillance efforts designed for Soviet deterrence and safeguarding NATO's interests in the Arctic, while also securing transatlantic supply lines critical for Western defense.

From *Wilson Center*, April 3 © 2024. Reprinted with permission. All rights reserved.

The end of the Cold War ushered in a new era of cooperation in the Arctic, characterized by dialogue and collaboration among Arctic states. NATO, in recognition of the changing dynamics of the region, adapted its approach to emphasize cooperation and engagement with regional partners, including NATO partners Finland and Sweden, and adopting the Norwegian motto "High North, Low Tension." One product of this era of renewed cooperation was the creation of the Arctic Environmental Protection Strategy, which laid the groundwork for the Arctic Council; the strategy is a multilateral agreement amongst Arctic states which addressed monitoring and assessments, environmental protections, emergency preparedness and response, and Arctic conservation. The new attitude towards an Arctic future materialized in participation in joint exercises and diplomatic cooperation with non-NATO members, through forums such as the NATO-Russia Council, further reflecting NATO's commitment to promoting wide-ranging stability and dialogue in the region.

## Adapting to a Changing Arctic Landscape

In recent years, the Arctic has once again emerged as a focal point for geopolitical competition and security challenges. Rapid climate change has led to the melting of ice caps, opening new shipping routes, access to natural resources, and potential military activity. Moreover, Russia's assertive actions, which include military buildup and exercises in the Arctic, have raised concerns about security and stability in the region. Diplomatic forums for dialogue, such as the operations of the Arctic Council, have been interrupted. These developments pose complex challenges that require a comprehensive and adaptive approach from NATO.

In response to these challenges, NATO has reaffirmed its commitment to the defense and security of the Arctic. The organization recognizes the need for a flexible and forward-looking approach that adapts to the evolving Arctic landscape. In his keynote address at the 10th Arctic Circle Assembly last year, Admiral Rob Bauer, Chair of the NATO Military Committee, unveiled Regional Plan North—one of several regional plans to be implemented by NATO as part of the organization's restructuring. Alongside the regional plan, NATO's new Arctic posture includes enhancing surveillance and reconnaissance capabilities, improving interoperability among member states, and strengthening partnerships with Arctic nations and multilateral organizations, such as the Arctic Council. This new posture was reiterated in the NATO Vilnius Summit Communique in July 2023, which stated "NATO and Allies will continue to undertake necessary, calibrated, and coordinated activities, including by exercising relevant plans." NATO's strategy in the Arctic emphasizes the importance of addressing both traditional and emerging security threats, ensuring the region remains stable and secure.

> **Rapid climate change has led to the melting of ice caps, opening new shipping routes, access to natural resources, and potential military activity.**

As NATO celebrates its 75th year, the enduring importance of the Arctic to the alliance underscores a dynamic partnership rooted in security, cooperation, and adaptation. Finland and Sweden joined the organization's ranks during the last year, bringing the total number of Arctic states in NATO to seven—the only outlying Arctic state being Russia. Despite contemporary geopolitical challenges, NATO's sustained involvement in the Arctic reflects a commitment to safeguarding its member states. As the Arctic continues to evolve, NATO will remain poised to adapt and respond, ensuring a peaceful and prosperous future for all members of the Arctic community.

## Print Citations

**CMS**: Donnelly, Elley. "No. 26 NATO in the Arctic: 75 Years of Security, Cooperation, and Adaptation." In *The Reference Shelf: U.S. National Debate Topic 2025–2026: Exploration & Development in the Arctic,* edited by Micah L. Issitt, 20–22. Amenia, NY: Grey House Publishing, 2025.

**MLA**: Donnelly, Elley. "No. 26 NATO in the Arctic: 75 Years of Security, Cooperation, and Adaptation." *The Reference Shelf: U.S. National Debate Topic 2025–2026: Exploration & Development in the Arctic,* edited by Micah L. Issitt, Grey House Publishing, 2025, pp. 20–22.

**APA**: Donnelly, E. (2025). No. 26 NATO in the Arctic: 75 years of security, cooperation, and adaptation. In M. L. Issitt (Ed.), *The reference shelf: U.S. national debate topic 2025–2026: Exploration & development in the Arctic* (pp. 20–22). Grey House Publishing. (Original work published 2024)

# Fulfilling the Promise of the Arctic Council

Paul Fuhs
*Northern Forum*, 2025

## Arctic Sustainable Development Principles

The Arctic, and our future, has become the subject of increased attention during the past decade, including by non-Arctic states and organizations. Almost everyone now has an Arctic strategy, or policy, everyone that is, except for a policy statement coming from the people who actually live here in the Arctic.

The closest thing to a Pan Arctic sustainable development policy is contained in the Ottawa Declaration, forming the Arctic Council, signed by the 8 Arctic nations in 1996. The Declaration is sound, but key elements of the Declaration have never been implemented.

The concept of the declaration is that the Arctic is a "special" place, requiring the *cooperation* of Arctic states and people, including a *"commitment to the well being of the inhabitants of the Arctic."*

The Council was established as a high level forum to: *"Provide a means for promoting cooperation, coordination and interaction among the Arctic States, with the involvement of the Arctic indigenous communities and other Arctic inhabitants on common arctic issues, in particular, issues of sustainable development and environmental protection in the Arctic."*

Issues of military and national security were strictly excluded.

This declaration was well received by the people of the Arctic and we viewed it as a solemn promise that our Arctic nation states made to each other, and more importantly, to the people of the Arctic, who well understand the value of Arctic cooperation. It was a recognition that the Arctic was so special that it warranted a whole separate, cooperative pan Arctic Council, in addition to the traditional diplomatic relations these nation states have with each other.

Over the 25 years of cooperation in the Arctic Council, tremendous accomplishments were made, including the sharing of Best Available Technology and Best Management practices for resource development, fisheries research and management, marine safety and emergency response, educational exchanges of all types, joint research and action on climate change, diplomatic and visa services, etc. All to the benefit of the *inhabitants* of the Arctic.

This Arctic cooperative work was never considered to be strategic from a military or national security point of view, and the Arctic was considered a 'zone of peace'. Nevertheless, despite our faithful exclusion of military issues in our work,

---

From *Northern Forum*, © 2025. Reprinted with permission. All rights reserved.

the Arctic Council has been militarized with a complete blockade against our cooperative work.

However, this blockade seems to be almost entirely symbolic, because by any sober assessment, this obstruction has had no impact whatsoever on the geopolitical dispute in Ukraine. What actually is lost, is the focus on benefits to the *inhabitants* of the Arctic, regardless of where we live. Our issues remain the same and the need for cooperation, given the impact of climate change, is stronger than ever.

Refocusing on these issues in a pan Arctic manner, will be one of the most important contributions that the Arctic Council can make to a post conflict world. There are already discussions in Norway, the current chair of the Council, to restart some of the pan Arctic work that can be done without extensive travel.

At a bare minimum, the immediate restart of pan Arctic cooperation should be included in any negotiated settlement of the Ukraine situation. That would he honoring the pledge that all Arctic nations made to each other and to the people of the Arctic in the formation of the Arctic Council

### Fulfilling the Work Plan of the Ottawa Declaration

Beyond the commitment to cooperation in general, the declaration laid out some specific work plans that have yet to be accomplished. One of the most important is the commitment to: *"adopt terms of reference for, and oversee and coordinate a sustainable development program."*

The declaration also affirmed a commitment to *"protection of the Arctic environment including...sustainable use of natural resources."*

This commitment has never been fulfilled, although it might be the most important, both as a statement of what 'sustainable development' means to the people of the Arctic, but also some specific actions necessary to implement them. A realistic sustainable development program can best be initiated by the sub national organizations around the Council including The Northern Forum (association of Arctic regional governments and states), The Arctic Mayor's Forum, The Arctic Economic Council, the Euro-Arctic Council, and the associated indigenous organizations across the Arctic.

### Overall Structure of the Arctic:

The demographic and physical structure of the Arctic includes a generally low and dispersed population base, surpluses of oil, gas, minerals, fiber, protein—fish and agricultural products, that can reliably and efficiently be exported to meet the needs of the heavily populated industrial Northern Hemisphere. Some high tech communication systems have also been developed. In addition, the Arctic provides unmatched tourism experiences for people from the overpopulated and environmentally degraded South. The Arctic is also characterized by a strong indigenous presence and heavy reliance on subsistence food security and culture. Therefore, high environmental standards for development are required. Air and marine transportation are critical to life in the Arctic, with the vast majority of imports and exports carried by maritime operations.

## Energy

Since response to 'climate change' is currently a driving force of Arctic policy, energy related issues are of particular importance. Along with basic human relations related to food gathering, energy has been the central organizing principle of civilization ever since humans learned to control fire. In the modern era, it has become indispensable. Arctic communities are dependent on it, yet we realize the impact of global climate change. How global energy policy is managed will have a critical impact on the future of the Arctic. An energy transition must be both practical and economic and must not further harm our Arctic communities.

Therefore, we declare these basic principles of an Arctic Sustainable Energy Policy.

- We recognize the impact of global climate change on the Arctic due to the consumption of fossil fuels by the 7 billion people of the non-arctic nations.
- We support the implementation of energy efficiencies to reduce consumption, the use of alternative fuels locally and globally and we recognize that traditional fuels will be necessary during a long transition period.
- We recognize the important contribution of the production of traditional fuels to Arctic economies, both in the private and public sector. Public revenues from production of traditional fuels have also often been used to finance renewable energy in the Arctic, which leads the world in alternative energy production.
- We support the production of hydrogen based alternative fuels using stranded energy in the Arctic, both for local consumption and for export.
- We support the development of environmentally responsible mining operations in the Arctic to produce the metals necessary for a transition to alternative fuels.
- We object to the selective discrimination against Arctic traditional fuels and mining operations by NGO's, governments and private sector banking and insurance companies.
- We support enhanced cooperation and prevention measures for shipment of fuels and cargo on the Northern Sea Route.

The Arctic regions are experiencing the most dramatic impacts of a warming world. Our indigenous people have been adapting to a warming planet for the past 10,000 years, at a time when ocean levels were 100 meters (330 feet) below their current levels. However, in the modern era, this warming has been accelerated by the consumption of fossil fuels.

The fact is, that halting all Arctic oil and gas development, as suggested by the European Union and various 'environmental' groups, won't result in even one drop less oil being burned. The Arctic produces about 10% of the world's petroleum resources, so an energy hungry world will just get its energy resources from somewhere else, often under lower environmental standards. Current proven world oil reserves indicate 53 years of supply.

The Arctic leads the way toward renewable energy: Iceland 100%, Greenland 81%, Alaska 30%, Canada 16%, Russia 21%, Denmark 32%, Norway 98%, Sweden 55%, Finland 25%. By comparison, the US is only 9% renewable, the OECD nations 10.5% and the non-Arctic EU 13.4%.

The campaign to blame the Arctic for climate change is particularly unwarranted. The World Wildlife Federation issued a report claiming, "Arctic nations produce 22% of CO2 Emissions". The only way they can claim this is by including the CO2 emissions of the ENTIRE United States at 13.4%. In fact, in Alaska, the Arctic part of the US, is only 0.006% of that number. A similar false calculation was used for Canada and Russia.

We also witnessed the false presentation of a dying polar bear by National Geographic under the title "This is what climate change looks like" even though no other polar bears in the area were in such condition. They eventually had to admit to this deception, but only after there had been 2.4 billion viewings of the image.

These false narratives and counterproductive policies around energy provide only harm for both Arctic residents and the consumers of the industrialized southern countries.

**Mining**
Mining is an important industry for Arctic economies and will also play a critical role in providing the necessary base and rare earth metals for batteries and drive motors for a non-carbon transition. For instance, an electric car requires 85 kg (193 pounds) of copper per vehicle. At a 3% recovery rate, three tons of copper ore must be mined for each electric vehicle. The Arctic can supply many of these critical mineral resources. Despite the need for these minerals, and the importance of mining projects to economies of the Arctic, we have seen the same patterns of lawsuits, protests, and discrimination by banks and insurance companies.

**Transportation**
Reciprocal port access and air landing agreements and diplomatic and visa services are critical to a well-functioning Arctic transportation system. Use of the Northern Sea Route can reduce $CO_2$ emissions by using a shorter route and can also provide efficient access to European and Asian markets for Arctic goods. The safety of this route can be enhanced by cooperative vessel tracking, monitoring and emergency response services.

**Subsistence**
Many Arctic cultures rely on local hunting, fishing, and gathering to provide for food security and to continue centuries old cultural traditions. We also recognize that modern subsistence tools and methods require a cash economy. Industrial

developments in the Arctic should be conducted in a manner to not harm these activities. Long term Indigenous and local knowledge should be respected and utilized in understanding a changing Arctic. We celebrate the continuation of indigenous cultural traditions and languages.

### Fisheries
As ocean waters warm, fish stocks are moving further north. Cooperative research and cooperative fisheries management agreements that have been disrupted will be necessary. In the Western Arctic, the Joint Norwegian-Russian Fisheries Commission continues its work for research and setting quotas and is a model for continued necessary cooperation.

### Tourism
Arctic tourism can provide economic benefits, but operators must be respectful of local cultures and provide real local benefits in their operations. Restrictions on visas and reciprocal flight operations must be lifted to allow travel for tourism.

### Forestry Operations
The Arctic zones can provide wood materials for construction and fiber for paper and other applications. Timber operations should meet the requirements of the Sustainable Forestry Protocol. As proposed at the recent COP26 deliberations in Glasgow, Arctic forests can provide a major carbon sink.

## Overall Environmental Protection
An additional work plan commitment of the Ottawa Declaration is to *"oversee and coordinate the programs established under the Arctic Environmental Protection Strategy on the Arctic Monitoring and Assessment Program (AMAP)Conservation of Arctic Flora and Fauna (CAFF), Protection of the Arctic Marine Environment (PAME), and Emergency Prevention, Prepared ness and Response (EPPR)."*

All of these working groups, none of which have any connection to military or national security issues, should be immediately restarted with all pan Arctic members of the Arctic Council. The Arctic issues that justified the formation of these working groups, all remain and must be addressed. The Arctic Coast Guard Forum should also be immediately reconstituted.

The agreement on the formation of the working group Cooperation on Marine Oil Pollution, Preparation and Response states that the Arctic nations are *"Aware of the Parties' obligation to protect the Arctic marine environment and mindful of the importance of precautionary measures to avoid oil pollution in the first instance."*

Recent technological advances include the ability to track and monitor vessel movements, to transmit navigation safety information directly to vessels, to direct rescue operations and to provide real time analysis of ice conditions and local weather. These should be applied across the Arctic to enhance prevention measures.

### Telecommunications

Arctic based corporations in the Arctic, particularly in Finland, led the world in developing mobile phone, server, and router developments. A trans Arctic fiber optic cable is currently being considered which can provide local service and will also be a critical redundancy to the world fiber optic system. This system and others such as satellite based Starlink, can provide high speed broadband access which is critical to Arctic communities for commerce, education, international cooperation, and health services. The goal should be to provide these services at costs similar to the industrialized South.

> **High environmental standards for development are required.**

### Arctic Investment and Finance

International banks and insurers have announced a policy of deliberate discrimination against Arctic oil and gas projects, simply because the projects are in the Arctic. This is the same historical racist 'redlining' practice used by banks in areas where certain people were considered to be substandard and this discrimination should be challenged through bank regulators. If this discrimination continues, we should consider forming an Arctic Development Bank with private and governmental financing agencies, that will evaluate projects based on their economic merits in conformance with the Arctic Investment Protocol adopted by the Arctic Economic Council.

### Education

An additional Ottawa Declaration commitment is to: *"disseminate information, encourage education and promote interest in Arctic-related issues."*

Preparing Arctic youth for their future is a critical function of sustainable development. Our young people must understand the realities of our economy and our environment and the related tradeoffs of risks and benefits of resource development and environmental policy. Preparedness for skilled labor and professions is essential and all jobs deserve equal respect. We pledge to develop an Arctic sustainable development curriculum which can be taught across the Arctic. We will also support student exchanges in education, culture, and sports.

### Diplomatic Relations

The use of broad-based sanctions and visa restrictions has been very harmful to the people of the Arctic and have restricted our ability to share best available technologies for environmental protection. The use of these sanctions has failed to provide tangible results and indicates a failure of diplomacy. The Arctic should continue to be viewed as an area of cooperation and peace, separate from other worldwide international conflicts. We recognize the importance of international cooperative councils such as the Arctic Economic Council, Barents Euro-Arctic Council, the Northern Forum and the Arctic Mayor's Forum which include the direct participation of local and regional governments along with indigenous organizations, in conjunction with national governments. While Arctic cooperation in these groups is now restricted by other non-Arctic events, these cooperative efforts should be allowed to immediately resume their activities once these other conflicts are resolved.

## A Peaceful and Productive Arctic

While we understand and respect our nation's efforts to maintain national security in the region, we continue to support the Arctic being treated as a peaceful zone of the world where we can cooperate with each other for a productive future for ourselves and future generations.

## Print Citations

**CMS**: Fuhs, Paul. "Fulfilling the Promise of the Arctic Council." In *The Reference Shelf: U.S. National Debate Topic 2025–2026: Exploration & Development in the Arctic*, edited by Micah L. Issitt, 23–29. Amenia, NY: Grey House Publishing, 2025.

**MLA**: Fuhs, Paul. "Fulfilling the Promise of the Arctic Council." *The Reference Shelf: U.S. National Debate Topic 2025–2026: Exploration & Development in the Arctic*, edited by Micah L. Issitt, Grey House Publishing, 2025, pp. 23–29.

**APA**: Fuhs, Paul. (2025). Fulfilling the promise of the Arctic Council. In M. L. Issitt (Ed.), *The reference shelf: U.S. national debate topic 2025–2026: Exploration & development in the Arctic* (pp. 23–29). Grey House Publishing. (Original work published 2025)

# The Golden Age of Offbeat Arctic Research

By Paul Bierman
*Undark*, September 6, 2024

In recent years, the Arctic has become a magnet for climate change anxiety, with scientists nervously monitoring the Greenland ice sheet for signs of melting and fretting over rampant environmental degradation. It wasn't always that way.

At the height of the Cold War in the 1950s, as the fear of nuclear Armageddon hung over American and Soviet citizens, idealistic scientists and engineers saw the vast Arctic region as a place of unlimited potential for creating a bold new future. Greenland emerged as the most tantalizing proving ground for their research.

Scientists and engineers working for and with the U.S. military cooked up a rash of audacious cold-region projects—some innovative, many spit-balled, and most quickly abandoned. They were the stuff of science fiction: disposing of nuclear waste by letting it melt through the ice; moving people, supplies, and missiles below the ice using subways, some perhaps atomic powered; testing hovercraft to zip over impassable crevasses; making furniture from a frozen mix of ice and soil; and even building a nuclear-powered city under the ice sheet.

Today, many of their ideas, and the fever dreams that spawned them, survive only in the yellowed pages and covers of magazines like "*REAL*: The exciting magazine for men" and dozens of obscure Army technical reports.

Karl and Bernhard Philberth, both physicists and ordained priests, thought Greenland's ice sheet the perfect repository for nuclear waste. Not all the waste—first they'd reprocess spent reactor fuel so that the long-lived nuclides would be recycled. The remaining, mostly short-lived radionuclides would be fused into glass or ceramic and surrounded by a few inches of lead for transport. They imagined several million radioactive medicine balls about 16 inches in diameter scattered over a small area of the ice sheet (about 300 square miles) far from the coast.

Because the balls were so radioactive, and thus warm, they would melt their way into the ice, each with the energy of a bit less than two dozen 100-watt incandescent light bulbs—a reasonable leap from Karl Philberth's expertise designing heated ice drills that worked by melting their way through glaciers. The hope was that by the time the ice carrying the balls emerged at the coast thousands or tens of thousands of years later, the radioactivity would have decayed away. One

From *Undark*, September 6 © 2024. Reprinted with permission. All rights reserved.

of the physicists later reported that the idea was shown to him by God, in a vision.

Of course, the plan had plenty of unknowns and led to heated discussion at scientific meetings when it was presented—what, for example, would happen if the balls got crushed or caught up in flows of meltwater near the base of the ice sheet. And would the radioactive balls warm the ice so much that the ice flowed faster at the base, speeding the balls' trip to the coast?

Logistical challenges, scientific doubt, and politics sunk the project. Producing millions of radioactive glass balls wasn't yet practical, and the Danes, who at the time controlled Greenland, were never keen on allowing nuclear waste disposal on what they saw as their island. Some skeptics even worried about climate change melting the ice. Nonetheless, the Philberths made visits to the ice sheet and published peer-reviewed scientific papers about their waste dream.

Arctic military imagination predates the Cold War. In 1943, that imagination spawned the Kee Bird—a mystical creature. An early description appears in a poem by A/C Warren M. Kniskern published in the Army's weekly magazine for enlisted men, *YANK*. The bird taunts men across the Arctic with its call "Kee Kee Keerist, but it's cold!" Its name was widely applied. Most well-known, a B-29 bomber named Kee Bird that took off from Alaska with a heading toward the North Pole, but then got badly lost and put down on a frozen Greenland lake in 1947 as it ran out of fuel. An ambitious plan to fly the nearly pristine plane off the ice in the mid-1990s was thwarted by fire. But the Kee bird lineage was by no means extinct.

In 1959, the *Detroit Free Press*, under the headline "The Crazy, Mixed-Up Keebird Can't Fly," reported that the Army was testing a new over-snow vehicle. This Keebird was not a flying machine but rather a snowmobile/tractor/airplane chimera that would cut travel time across the ice sheet by a factor of 10 or more. Unlike similar but utilitarian contraptions of the 1930s, developed in the central plains of North America and Russia and equipped with short skis, boxy bodies, and propellors that pushed them along, this new single-propped version was built for sheer speed.

The prototype hit 40 miles per hour at the Army's testing facility in Houghton, Michigan, thanks to the "almost friction-proof" Teflon coating on its 25-foot-long skis and a 300-horsepower airplane engine that spun the propeller. The goal was for the machine to hit a hundred miles per hour but after several failed tests, and a few technical publications, it warranted only the one syndicated newspaper article written by Jean Hanmer Pearson, who was a military pilot in World War II before she became a journalist and one of the first women to set foot on the South Pole. The Soviet version, known as an "airsleigh," was short, stout, and armed with weapons for Arctic combat. There's no record the Army's Keebird carrying weapons.

In 1964, the Army tested a distant relative of the Keebird in Greenland. The Carabao, which floated over the ground and over water or snow on a cushion of air, was developed by Bell Aerosystems Company and had been previously tested in tropical locales, including southern Florida. It carried two men and 1,000 pounds of cargo, and had a top speed of 60 miles per hour. The air cushion vehicle skimmed over crevasses but was grounded by even moderate winds, an all-too-common occurrence on the ice sheet.

Another problem: The craft went uphill fine, but going downhill was another matter because it had no brakes. Unsurprisingly, the Carabao—its namesake a Philippine water buffalo—proved to be unsuited for ice travel despite the claim that: "All this is no mere pipe-dream following an overdose of science-fiction. The acknowledged experts are thinking hard about the future use of hovercraft in Polar travel." Despite all the hard thinking, hovercraft have yet to catch on and are still rarely used for Arctic travel and research.

In 1956, *Colliers*, a weekly magazine once read by millions of Americans, published an article titled "Subways Under the Icecap." It was a sensationalized report of Army activities in Greenland and opened with a photograph of an enlisted soldier holding a pick. Behind him, a 250-foot tunnel, mostly excavated by hand and lit only by lanterns, probed the Greenland ice sheet. *Colliers* included a simple map and a stylistic cut-away showing an imaginary rail line slicing across northwestern Greenland. But the Army's ice tunnels ended only about a thousand feet from where they started—doomed by the fragility of their icy walls, which crept inward up to several feet each year, closing the tunnels like a healing wound. The subway never happened.

That didn't stop the Army from proposing Project Iceworm—a top-secret plan that might represent peak weirdness. A network of tunnels would crisscross northern Greenland over an area about the size of Alabama. Hundreds of missiles, topped with nuclear warheads, would roll through the tunnels on trains, pop up at firing points, and if needed, respond to Soviet aggression by many annihilating many Eastern Block targets. Greenland was much closer to Europe than North America, allowing a prompt strategic response, and the snow provided cover and blast protection. Iceworm would be a giant under-snow shell game of sorts, which the Army would power using portable nuclear reactors.

Except it wasn't a game. The Army hired the Spur and Siding Constructors Company of Detroit, Michigan, to scope out and price the rail project. A 1965 report, complete with maps of stations and sidings where trains would sit when not in use, concluded that contractors could build a railroad stretching 22 miles over land and 138 miles inside the ice sheet for a mere $47 million (or roughly $470 million today). The company suggested studying nuclear-powered locomotives because they reduced the risk of heat from diesel engines melting the frozen tunnels. Never mind that no one had ever built a nuclear locomotive or run rails through tunnels crossing constantly shifting crevasses.

But in the end, Iceworm amounted only to a single railcar, 1,300 feet of track, and an abandoned military truck on railroad wheels.

The split personality of Arctic permafrost frustrated Army engineers. When frozen in the winter, it was stable but difficult to excavate. But in the summer, under the warmth of 24-hour sunshine, the top foot or two of soil melted, creating an impassable quagmire for people and vehicles. When the permafrost under airstrips melted, the pavement buckled, and the resulting potholes could damage landing gear. The military responded by painting Arctic runways white to reflect the constant summer sunshine and keep the underlying permafrost cool—a potentially good idea grounded in physics that was stymied by the fact that the paint reduced the braking ability of planes.

The military engineers, ever optimistic, put a more positive spin on permafrost. Trying to use native materials in the Arctic, where transportation costs were exceptionally high, they made a synthetic version of permafrost that they nicknamed permacrete—a mashup of the words permafrost and concrete. First, they mixed the optimal amount of water and dry soil. Then, after allowing the mix to freeze solid in molds, they made beams, bricks, tunnel linings, and even a chair. But permacrete never caught on as a building material, likely because one warm day was all it would take to turn even the most robust construction project into a puddle of mud.

The Army's most ambitious Arctic dream actually came true. In 1959, engineers began building Camp Century, known by many as the City Under the Ice. A 138-mile ice road led to the camp that was about 100 miles inland from the edge of the ice sheet. Almost a vertical mile of ice separated the camp from the rock and soil below.

Camp Century contained several dozen massive trenches, one more than a thousand feet long, all carved into the ice sheet by giant snowplows and then covered with metal arches and more snow. Inside were heated bunkrooms for several hundred men, a mess hall, and a portable nuclear power plant. The first of its kind, the reactor provided unlimited hot showers and plenty of electrical power.

The camp was ephemeral. In less than a decade, flowing ice crushed Century—but not before scientists and engineers drilled the first deep ice core that eventually penetrated the full thickness of Greenland's ice sheet. In 1966, the last season the Army occupied Camp Century, drillers recovered more than 11 feet of frozen soil from beneath the ice—another first.

Little studied, the Camp Century soil vanished in 1993, but was rediscovered by Danish scientists in the late 2010s, safely frozen in Copenhagen. Samples revealed that the soil contained abundant plant and insect fossils, unambiguous evidence that large parts of Greenland were free of ice some 400,000 years ago, when the Earth was about the same temperature as today but had almost 30 percent less carbon dioxide in the atmosphere.

> **Scientists and engineers cooked up a rash of audacious cold-region projects: disposing of nuclear waste by letting it melt through the ice; moving people, supplies, and missiles below the ice using subways; testing hovercraft to zip over impassable crevasses; making furniture from a frozen mix of ice and soil; and building a nuclear-powered city under the ice sheet.**

In [a] half century or so since the demise of Camp Century, global warming has begun melting large amounts of Greenland's ice. The past 10 years are the warmest on record, and the ice sheet is shrinking a bit more every year. That's science, not fiction, and a world away from the heady optimism of the Cold War dreamers who once envisioned a future embedded in ice. (*Update*: A previous version of this piece stated that Camp Century soil was rediscovered by Danish scientists in 2018. The piece has been updated to reflect differing recollections among researchers.)

**Print Citations**

**CMS**: Bierman, Paul. "The Golden Age of Offbeat Arctic Research." In *The Reference Shelf: U.S. National Debate Topic 2025–2026: Exploration & Development in the Arctic*, edited by Micah L. Issitt, 30–34. Amenia, NY: Grey House Publishing, 2025.

**MLA**: Bierman, Paul. "The Golden Age of Offbeat Arctic Research." *The Reference Shelf: U.S. National Topic 2025–2026: Exploration & Development in the Arctic,* edited by Micah L. Issitt, Grey House Publishing, 2025, pp. 30–34.

**APA**: Bierman, P. (2025). The golden age of offbeat Arctic research. In M. L. Issitt (Ed.), *The reference shelf: U.S. national debate topic 2025–2026: Exploration & development in the Arctic* (pp. 30–34). Grey House Publishing. (Original work published 2024)

# 2
# Indigenous Interests

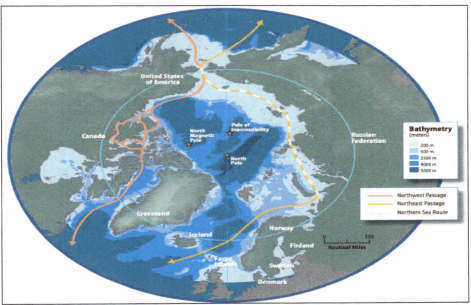

Arctic region map (2009) detailing shipping routes Northeast Passage, Northern Sea Route, and Northwest Passage. Map by Susie Harder, Arctic Council, via Wikimedia. [Public domain.]

# Independent Action in the Arctic

The debate about the Arctic often focuses on government action, inaction, and mismanagement, but there are many different players active in the Arctic region whose activities are less often featured in this debate. This includes nongovernmental organizations (NGOs), which operate independently of any government, but may work with governments to achieve certain goals. The Arctic debate also involves the activities of a number of Indigenous groups focused on representing and promoting the views and welfare of the Indigenous residents of the Arctic region.

## Types of NGOs Active in the Arctic

The central debate over the management of the Arctic region is about balancing the desire, on the part of corporations and governments, to harvest valuable resources from Arctic lands and to use these lands and surrounding seas as military and economic properties, against the effort to protect and preserve Arctic wildlife and the Indigenous people who have been living in these unique environments for thousands of years. Some people feel that harvesting resources is more important, while others feel that the preservation of ecosystems, wildlife, and Indigenous communities is more important, and others are trying to balance the scales between these goals, looking for ways to preserve and harvest simultaneously. This goal, of compromising between economic and military interest and preservation/conservation has been difficult to achieve, and intergovernmental cooperation, as through the Arctic Council, has been essential in this effort, especially as the Arctic Council has worked with and invited input from NGOs and has incorporated Indigenous perspectives into various projects. However, with the United States turning away from conservation and international cooperation, it remains unclear how the Arctic Council will function, and this means that the more narrowly focused efforts of NGOs might become even more important in the Arctic moving forward and in shaping the debate over Arctic resources among participating governments.

One of the essential debates in the realm of the Arctic concerns the fate of the wildlife that evolved to live in this region of the world. Arctic animals and plants face numerous threats, including pollution and habitat destruction from drilling and fossil fuel harvesting, and from climate change, the most pressing and dramatic threat to the future of Arctic ecosystems. There are a number of NGOs active in the Arctic focused entirely on the preservation of ecosystems and wildlife and these NGOs fund and support research that helps us better understand how economic and military activities in this region will change the ecosystem and how this might impact humanity in the future as well.

Among the NGOs with a presence in the Arctic, none are better known than the World Wildlife Fund (WWF). Founded in 1961 by independent Swiss conservationists the WWF has achieved great visibility and international recognition over the course of the twentieth and twenty-first centuries for their efforts to support wildlife conservation and to address issues where human activities negatively impact ecosystems and animal species. The WWF is one of the largest NGOs, active in more than 100 countries and operating through thousands of different environmental projects in specific areas. In the Arctic, the WWF has mounted a very visible campaign to save the polar bear (*Ursus maritimus*) from extinction. The polar bear, a globally famous and famously charismatic species, makes a good symbolic focal point for Arctic wildlife more generally, but there are actually many different species being effected by changing conditions and the loss of habitat, including the Arctic wolf, a number of Arctic seal species, the Arctic fox, and the Pacific salmon, which is an important food crop for humans in addition to a key species in a complex set of ecosystems. The Arctic seas, meanwhile, not only support species that live in those waters year-round, but form an important part of the migratory pathway for a number of species, including whales and seals and several species of shark.[1]

Another well-known NGO very active in the campaign for Arctic preservation is Greenpeace, an independent campaign network that was started by environmental activists in Canada in the early 1970s. Greenpeace also supports efforts to conserve wildlife, but their projects range into other areas as well, including nuclear disarmament and antiwar activism. Greenpeace's "Save the Arctic" campaign, which began in 2012, calls specifically for a sanctuary to be established in the area around the North Pole. In addition to petitioning for government action, Greenpeace activists have taken their campaign directly to the oil companies actively drilling in the Arctic, staging protests directly at drilling operations. In 2013, more than thirty Greenpeace activists were taken captive by the Russian Coast Guard. The captives were freed after three months after the Russian government declared a general amnesty for all protestors involved.[2]

Beyond conservation, there are other NGOs active in the Arctic promoting a more general focus on scientific discovery and understanding. These including the International Arctic Science Committee (IASC), an NGO founded in 1990 with representatives of all eight Arctic countries. By the 2020s, the IASC had grown to promote programs on Arctic research and science in twenty-four different countries. The organization's prime goal is to encourage international cooperation and participating in joint programs, and IASC research programs are investigating things like atmospheric science, Arctic marine habitats, and how human activities are changing the Arctic environment.[3] Another well-known NGO supporting scientific research in the Arctic is ArcticNet, which is a network of Canadian organizations collectively contributing to the study of climate change in the Arctic region. ArcticNet creates and supports efforts to bring students and researchers to the Arctic and to facilitate new research programs with climate relevance.[4]

Beyond the sciences, there are NGOs more directly involved in trying to shape governmental policy and economic policy in the region. One example is the Arctic Institute: Center for Circumpolar Security Studies. This is one of the newer NGOs, established in 2011, with a focus on security and economic issues in the region. The think tank, headquartered in Washington, D.C., but involving an international panel of advisors and directors, supports research on Arctic economic programs, including the "blue economy," as well as how Arctic nations can manage climate change. They also support studies on population shifts in the Arctic, urban design and planning, and governance of both disputed territories and the ocean. The Arctic Institute routinely ranks among the highest-rated and most-respected think tank dealing with Arctic issues and has won praise from politicians, scientists, and researchers, for attempting to bridge the gaps between scientific data and public policy. Rather than advocating directly for specific initiatives, the Arctic Institute and similar NGOs are more focused on the human ecosystem connection and how this relates to governmental and economic policy. In 2025, Arctic Institute researchers and journalists have focused on following the developments of military buildup in the Arctic and how this is connected both to increasing Russian military aggression and the loss of sea ice, which has made military activity in the Arctic realm more accessible.[5]

## Indigenous Voices and Perspectives

The Indigenous inhabitants of the Arctic nations have endured many years of marginalization and exploitation and now, in the twenty-first century, the activities of the industrialized world not only threaten Arctic ways of life, but the integrity of the landscape in which their ancestors established the roots of their cultures. In American popular culture, the indigenous peoples of the Arctic are often imagined as representing one relatively homogenous cultural block, but this is an inaccurate perception. There are more than forty different ethnic groups living in the Arctic region, together representing a full 10 percent of the population of these countries.

Among the best-known Arctic cultures are the Sámi (sometimes spelled "Sáami") people, who live in areas within the countries of Finland, Norway, Russia, and Sweden. Once known as the "Lapps" or "Laplanders," that term is now considered offensive and the Sámi people have shifted to utilizing their own terminology. Famous for their reindeer herding communities, other Sámi people are more focused on fishing, sheep agriculture, and fur harvesting. Some of these ecological roles are legally allowed only for Sámi people in Norway, in an effort to balance the preservation of traditional cultural modalities with the effort to preserve wildlife and species. The Sámi people are struggling to maintain traditional lifestyles under pressure from many different sources. For one thing, there have been numerous conflicts between the Sámi and European peoples in the region over ownership of land and territory, especially territories desires for natural resource exploitation.[6] A number of NGOs in the region have programs involving Sámi rights and autonomy, but the Sámi also have their own NGO, the Sámiráddi, or "Saami Council," which involves nine different Sámi organizations located in Finland, Norway, Sweden, and Russia. Founded in 1956, the Saami Council is one of the oldest Indigenous organizations in the world and they have worked with both governments and other NGOs to promote Sámi viewpoints and interests as the Arctic has changed.[7]

Another of the largest and most influential Indigenous populations in the Arctic are the Inuit, a large group of culturally and historically related Indigenous people living across the Arctic, with major population centers in Greenland, Labrador, Quebec, Nunavit, the Northwest Territories, Yukon, Alaska, and Russia. Canada's Inuit population forms a major population center for the Inuit and more than 90 percent of the inhabitants of this sparsely inhabited territory (officially part of Canada's Quebec province) are Inuit in descent. In Greenland, the ancestors of the original Inuit, known as the Kalaallit, are descended from the migration of Indigenous inhabitants through Canada. Living in a former Danish province, Kalaallit in Greenland are Danish citizens by birth and citizens of the European Union (EU). In the United States, the closely related Iñupiat live in parts of Alaska, and in Russia, there are Iñupiat people that once resided on the Bering Strait islands controlled by Russia. At one point, all the Indigenous people who might be called Inuit were known as "Eskimos" in United States culture, a term that has been integrated into pop culture in a number of ways. Many of those who identify as Inuk (the singular of Inuit) reject the label "Eskimo" as an artifact of colonialism. A closely related group known as the Yupik, Indigenous to parts of Alaska and Russia, and actually consisting of at least three distinct groups, are closely related to the Inuit people, but some still utilize the term "Eskimo" in self-identification.

Inuit and Yupik communities face different challenges depending on what nation currently controls the territory in which their communities reside. Canada, Norway, Sweden, and Finland have programs in place to provide some degree of self-representation and autonomy to Inuit people within those territories, and these programs function, to some degree, to protect Indigenous rights and freedoms, and to help Indigenous peoples to preserve their own cultures, but these remnants of colonialism are likewise imperfect and often flawed.

In Canada, for instance, Inuit people face challenges related to cultural, social, and economic marginalization. Years of racist policy and law have left Inuit and other Indigenous communities deeply marginalized and modern Indigenous communities struggle with high levels of income and resource disparity, problems made worse by climate change and shifting environmental patterns. In the United States, by contrast, there has been far less focus on elevating or embracing Indigenous perspectives in policy. This is true in the continental United States, where Indigenous Americans are still marginalized and face prejudicial policies, and in Alaska, where the same is true of the nation's Inuit and Yupik residents.

The island nation of Greenland is the only place in the world where Inuit dominate in the population, comprising some 89 percent of the island. Greenland was once a territory of Denmark, until 1953, and has since transitioned to independence. The 2008 Greenlandic self-government referendum transferred much of the remaining authority from Denmark to Greenland, though the Danish government continues to be involved, managing citizenship, some economic policies, and foreign affairs. Some Inuit living in Greenland want even more autonomy and separation from the Danish government, while others favor maintaining strong ties to Denmark, which provides some advantages in international influence and grants Greenland citizens Danish and EU citizenship as well. Among the threats facing Inuit in Greenland include the dramatic rise in temperatures that is rapidly changing the environment of the country, and the ongoing effort to promote national welfare while addressing demands for Greenland resources. Donald Trump has suggested that Greenland should be under US control and has advanced the false claim that Greenland's people favor his eventuality, despite polls showing that the vast majority of Greenland residents would oppose US colonization. So, Greenland's Inuit population, in addition to facing problems related to resource insecurity and the ongoing and worsening impact of climate change, are now forced to grapple with the threat of colonization from a nation once seen as an important and valued ally.

Inuit people have government positions in many of the countries where Inuit people live, but they are also marginalized minorities throughout much of their range, such as in the United States, where Inuit in Alaska have been repeatedly and continually sidelined in favor of economic development and resource exploitation. There are a number of NGOs focused on Inuit welfare who have also played a role in the Arctic debate. One of the oldest of these focused on Inuit rights welfare and perspectives is the Inuit Circumpolar Council, an NGO founded in 1977 and headquartered in Canada that represents more than 180,000 Inuit and Yupik people in Canada, Alaska, Greenland, and the Chukchi Peninsula of Russia. The ICC is considered a special consulting organization to the United Nations, and hold annual conferences on Inuit and Yupik issues.

## A Combination of Voices

The Arctic region has become a focal point in 2020s politics, in part, because the melting of the sea ice has increased accessibility. This might mean greater utility as military properties, or might mean enhanced access to natural resources, but the management of the Arctic and its resources cannot be about military advantage and economic opportunity alone without sacrificing much of what makes the Arctic unique and further marginalizing the Indigenous people whose cultures and sometimes lifestyles are linked to the landscape and resources under pressure from multiple threats. Over the years, Indigenous groups and NGOs have played an important role in shaping attitudes and building knowledge about the Arctic and the people who live in the Arctic nations, and these organizations are continuing to support research, public and private alliances, and activism aimed at addressing threats to Arctic communities and environments, while also helping citizens of the Arctic and around the world, to understand the stakes in this ongoing debate and contest for control and autonomy in the Arctic region.

## Works Used

"About." *International Arctic Science Committee (IASC)*, 2025, www.iasc.info/about.

"About the Saami Council." *Saamicouncil.net*, 2025, www.saamicouncil.net/en/the-saami-council.

"About Us." *Arctic Institute*, 2025, www.thearcticinstitute.org/about-us/.

"The Indigenous World 2024: Sápmi." *International Working Group for Indigenous Affairs (IWGIA)*, 2025.

"Protect the Arctic." *Greenpeace*, www.greenpeace.org/usa/protect-the-arctic/.

"Working Together in a Changing Canadian Arctic." *ArcticNet*, 2025, arcticnet.ca/.

"WWF Global Arctic Programme." *Arctic World Wildlife Fund (WWF)*, 2025, www.arcticwwf.org/.

## Notes

1. "WWF Global Arctic Programme," *Arctic World Wildlife Fund (WWF)*.
2. "Protect the Arctic," *Greenpeace*.
3. "About," *International Arctic Science Committee (IASC)*.
4. "Working Together in a Changing Canadian Arctic." *ArcticNet*.
5. "About Us," *The Arctic Institute*.
6. "The Indigenous World 2024: Sápmi," *International Working Group for Indigenous Affairs (IWGIA)*.
7. "About the Saami Council," *Saamicouncil*.

# Many People in the Arctic Are Staying Put Despite Climate Change, Study Reports

By Katie Bohn
*Penn State University Agricultural Sciences*, May 10, 2024

University Park, PA.—Temperatures are rising rapidly in the Arctic, raising questions about how communities are coping in the shifting climate. A team led by Penn State researchers reviewed studies from the past 30 years to examine whether these challenges are causing people to migrate out of the area—or if, and why, they're deciding to stay.

The researchers, who published the findings in Regional Environmental Change, found little evidence of individuals or households migrating out of the polar regions of Alaska and northern Canada as a result of climate change. Factors such as family, culture and a sense of community led to people deciding to stay, even in the face of climate-related challenges.

However, the researchers did find evidence of whole communities relocating when climate change led to deteriorating conditions. For example, the Chevak Native Village in Alaska already has relocated. In another example, the village of Newtok—a Yup'ik community in Alaska—has spent millions of dollars on relocation efforts, which already have taken more than 30 years and are not yet complete.

"Arctic communities under environmental threats are forced to relocate because flooding, erosion and storms are destroying their homes and infrastructure," said the paper's lead author, Guangqing Chi, professor of rural sociology, demography and public health sciences in Penn State's College of Agricultural Sciences. "Community relocation from climate-related environmental changes is a widely considered option in Alaska, but it is an expensive process."

As of 2022, the researchers said, 144 of the 229 Alaska native tribes were under environmental threats, including 29 communities experiencing significant erosion, 38 communities facing significant flooding and 35 having problems with thawing permafrost.

Of those, 15 are exploring relocation, which includes moving housing and public infrastructure. But many communities that are facing environmental threats cannot meet the requirements of federal disaster mitigation programs and are ineligible for disaster funding.

Chi, who is also a Social Science Research Institute co-funded faculty member and the director of the institute's Computational and Spatial Analysis Core,

From *Penn State University Agricultural Sciences*, May 10 © 2024. Reprinted with permission. All rights reserved.

said the findings highlight the need for further research, as well as recommendations for tackling the unique challenges of studying this topic.

"Many of these communities are facing multiple challenges, including thawing permafrost, declining sea ice cover, coastal erosion and extreme storms," said Ann Tickamyer, a co-author and professor emerita of rural sociology and demography at Penn State. "As critical climate tipping points are reached, threats to these communities' viability, health and livelihoods will only increase."

But despite temperatures in the Arctic rising four times more quickly than in lower latitudes, the researchers said the region largely is absent from studies and debates on how climate change is fueling migration.

> **Individuals and households that relocated out of the Arctic tended to do so because of factors that influence migration everywhere, such as jobs, education and health care.**

"This is an important gap because of the severity of Arctic climate change impacts and the regional predominance of Indigenous communities—many of which have already been negatively impacted by centuries of racism, cultural loss and political disenfranchisement, especially in Alaska," Chi said.

To arrive at their conclusions, the researchers examined peer-reviewed studies on the factors contributing to migration in the Arctic—including those related and not related to the climate and environment.

After their analysis, the researchers found no evidence of individuals or households moving because of shifts in the climate. Instead, individuals and households that relocated out of the Arctic tended to do so because of factors that influence migration everywhere, such as jobs, education and health care.

"For example, a study of 43 Alaska towns and villages—which covered places most threatened by climate-linked erosion and flooding—found no indication of enhanced out-migration between 1990 and 2014 compared to places without climate risk," said co-author Shuai Zhou, a former doctoral student in rural sociology and demography at Penn State and currently a postdoctoral associate at Cornell University.

Chi said the review is part of the Pursuing Opportunities for Long-Term Arctic Resilience for Infrastructure and Society (POLARIS) project. The initiative includes experts from different disciplines and institutions working toward helping Arctic Indigenous populations adapt and become resilient to environmental changes.

The team currently is conducting surveys and in-depth interviews to understand climate change impacts on community wellbeing and to seek indigenous knowledge in dealing with climate extremes and hazards.

In the future, the researchers said they plan to conduct new research to better understand the challenges faced by Arctic communities, as well as the potential solutions.

**Print Citations**

**CMS**: Bohn, Katie. "Many People in the Arctic Are Staying Put Despite Climate Change, Study Reports." In *The Reference Shelf: U.S. National Debate Topic 2025–2026: Exploration & Development in the Arctic,* edited by Micah L. Issitt, 43–45. Amenia, NY: Grey House Publishing, 2025.

**MLA**: Bohn, Katie. "Many People in the Arctic Are Staying Put Despite Climate Change, Study Reports." *The Reference Shelf: U.S. National Debate Topic 2025–2026: Exploration & Development in the Arctic,* edited by Micah L. Issitt, Grey House Publishing, 2025, pp. 43–45.

**APA**: Bohn, K. (2025). Many people in the Arctic are staying put despite climate change, study reports. In M. L. Issitt (Ed.), *The reference shelf: U.S. national debate topic 2025–2026: Exploration & development in the Arctic* (pp. 43–45). Grey House Publishing. (Original work published 2024)

# The True Cost of Mining in the Canadian Arctic

By Henry Harrison
*The Circle (WWF)*, April 2023

Operated by a Canadian mining company called Baffinland, the Mary River Mine is one of the world's richest reserves of high-grade iron ore. Located on the northern end of Baffin Island in Nunavut, the mine produces 4.2 million tonnes of ore each year. In 2024, that amount will increase to 6 million.

Last fall, the Canadian government rejected Baffinland's proposal to further expand its operations. The company's plans would have more than doubled the mine's output and seen a 110 km railway built to service the nearby port. Although the government's decision was a victory for many Inuit groups in the region—who fought the proposal—the 1,700 residents of Pond Inlet are still living with the consequences of having an iron mine as a neighbour. Enookie Inuarak is one of them. He's the former vice chairperson of the Mittimatalik Hunters & Trappers Organization, one of the groups that fought the expansion plan. He spoke to *The Circle* about the impact the mine has had on his community since it opened in 2015.

How would you describe Pond Inlet?

It's one of the most northern communities in Nunavut, above the Arctic Circle. Right now, it's the dark season. We haven't seen the sun for at least two or three weeks. This area is a highway for marine wildlife during spring and summer —and even in winter. We see species like bowhead whales, narwhals, belugas, polar bears, different kinds of seals and a lot of birds. We are still very active in harvesting for subsistence use. We depend on it for food and clothing. It's a very important part of our diet, especially narwhals, different kinds of seals, caribou and Arctic char.

When Baffinland opened the mine in 2015, how did it affect your community?

It's so different now. First off, we started noticing the marine wildlife not being around as much as before. And we ended up having to travel longer and farther to hunt and spend more on gas. We also had to start spending more on store-bought food.

But less marine wildlife was just one of the first impacts we noticed. We're harvesting less char now during the summer. We see fewer seals and narwhals, and even birds, now. There are constantly ships coming and going, anywhere from two to five a day, and there's constant noise pollution. Even before there

From *The Circle (WWF)*, April © 2023. Reprinted with permission. All rights reserved.

were any mining ships, when any kind of ship passed by, we used to notice marine wildlife being scared and moving away. Now we see constant ships, and I guess it's the noise pollution they're avoiding.

If those species were to disappear, what would it mean to your way of life and your community?

Without these species, we would also lose our harvesters and hunters. The income that comes with those activities would be lost, so the harvesters and hunters would have to look for different ways to make a living. And that would be similar to losing farmers in southern areas: imagine no more steak, eggs, bacon and so on. It would be a catastrophe, and countless centuries-old traditions would be gone.

It would definitely affect us, health-wise, because the food we harvest now is very healthy. For instance, with narwhals we get a lot of the vitamin C we need, and without them, we would get a lot less. If we were to rely on only store-bought food—which is shipped to Pond Inlet just once a year—it would have a major impact.

What role did you have in fighting the mine's expansion plans?

Well, the Mittimatalik Hunters & Trappers Organization represents the Inuit of the community as well as Inuit hunting rights and the environment. In one of our annual general meetings, someone put forward a motion not to support the expansion because of the impacts we were already seeing. I think we had a big role in preventing the expansion—we were the voice of the community, we were part of the hearings, and we also went down to Ottawa to lobby. We talked about the reality of what we were seeing and the changes we had seen.

The mining company kept saying there wouldn't be impacts, but that didn't make any sense at all because we see the impacts already. I think we definitely had some role in the government's decision. We were very happy that the government listened to us.

> **The impacts to the Inuit way of hunting and sustaining our culture could be irreversible.**

What do you see for the future? Can your community and the species that live there coexist with this mine?

Well, hopefully the marine wildlife can adapt somewhat to this new constant ship traffic. But the way we see things now, it's kind of difficult. Like, some families here have gone out on the land hunting and come back empty-handed. It's sad. And it hurts. So, I don't know how it's going to be.

And now, with the increase in production to 6 million tonnes next year, there is going to be more shipping and a longer shipping season. Already this year, we saw ships passing by when ice was forming, and the hunters noticed that the seals were completely gone until the ships stopped coming. Hopefully, the mining company can listen to Inuit and work with the governments and different organizations. Because the impacts to the Inuit way of hunting and sustaining our culture could be irreversible. And if this keeps happening, what do we get in return? We barely get anything in return.

It could be devastating in the long term. What if I can't teach my children our ways anymore, and my children's children don't learn what I learned from my father? That is what I am scared to think about.

## Print Citations

**CMS**: Harrison, Henry. "The True Cost of Mining in the Canadian Arctic." In *The Reference Shelf: U.S. National Debate Topic 2025–2026: Exploration & Development in the Arctic,* edited by Micah L. Issitt, 46–48. Amenia, NY: Grey House Publishing, 2025.

**MLA**: Harrison, Henry. "The True Cost of Mining in the Canadian Arctic." *The Reference Shelf: U.S. National Debate Topic 2025–2026: Exploration & Development in the Arctic,* edited by Micah L. Issitt, Grey House Publishing, 2025, pp. 46–48.

**APA**: Harrison, H. (2025). The true cost of mining in the Canadian Arctic. In M. L. Issitt (Ed.), *The reference shelf: U.S. national debate topic 2025–2026: Exploration & development in the Arctic* (pp. 46–48). Grey House Publishing. (Original work published 2023)

# Valuing Indigenous Knowledge in Permafrost Research

By Meral Jamal
*Undark*, January 10, 2024

Over the last two years, Emma Street has taken trips to Canada's North to places such as Tuktoyaktuk, a hamlet of less than a thousand people in the Northwest Territories, and Ulukhaktok, a small community on the west coast of Victoria Island. In these remote towns, Street, a Ph.D. student at the University of Victoria, has been meeting with Indigenous community members to learn about the Arctic's changing landscape and how it is affecting their way of life.

"This is people's lives and livelihoods and cultural connection," said Street.

In March, she interviewed Irma and Ernie Francis, a Gwich'in couple who live in Inuvik, a town located about 120 miles north of the Arctic circle. Along the Mackenzie River, they saw houses sinking, the ground eroding beneath them. Community members shared how they've had to relocate due to the damage caused to their houses.

"It's just unbelievable," Irma Francis said. "I've never seen it like this in my 57 years living in Inuvik."

Permafrost—ground that is continually frozen for at least two years—underlies anywhere from 15 to 25 percent of the Northern Hemisphere (depending on calculation methods), and stores up to an estimated 1.6 trillion metric tons of organic carbon in the region, twice the amount currently held in Earth's atmosphere. Scientists know that the Arctic, which contains the majority of the planet's permafrost, is warming four times faster than the rest of the world, and that as temperatures rise, permafrost will release carbon, exacerbating the effects of climate change.

But according to Patrick Murphy, a field research technician with the Woodwell Climate Research Center in Massachusetts, it's unclear how quickly that degradation is taking place. "Natural emissions from the permafrost are unknown, in the sense that we have a few measurements," he said. Monitoring stations "have existed for decades at this point—but only in a few places."

Murphy and his colleague Kyle Arndt want to change that, allowing updated climate models to account for permafrost emissions and make more accurate predictions. They are part of a recent, multi-institutional project called Permafrost Pathways, which is quantifying permafrost degradation across the Arctic boreal region and using the results to guide more realistic climate policy.

From *Undark*, January 10 © 2024. Reprinted with permission. All rights reserved.

A key component to the project's success, Arndt said, is involving northern Indigenous communities—the people who will feel the effects of thawing permafrost especially acutely. "We want to be involved and listening to their concerns as well," he wrote in an email to *Undark*.

Street, meanwhile, is part of a similar research initiative called PermafrostNet, a sprawling network of scholars, researchers, and government agencies, largely based in Canada, studying the effects of permafrost thaw on communities and developing measures to adapt. Her research deals with how melting permafrost is not only affecting infrastructure, but also Indigenous culture and customs.

Indigenous knowledge—which is rooted in the worldviews held by Indigenous peoples and their lived experiences—has, until recently, rarely been prioritized in scientific research, said Pascale Roy-Léveillée, another investigator with PermafrostNet, as well as an associate professor of geography at Université Laval in Quebec. For example, a 2021 literature review found that only five studies across a variety of disciplines used the terms "Indigenous Knowledge," "Traditional Ecological Knowledge," or "Indigenous Ecological Knowledge" in 1990.

That is beginning to change: In 2018, the number of such studies had increased to 1,404. The increase coincides with calls from within the scientific community for improved engagement with Indigenous communities. And many Indigenous communities now require that scientists undergo a local review process to ensure responsible community engagement.

"I think Indigenous knowledge can be very powerful for getting attention on work and is a huge largely untapped resource and knowledge base," said Arndt.

But such research still faces challenges, including time and funding constraints. Meanwhile, relatively few members of academic research teams are Indigenous themselves, and some climate scientists have warned against extracting local knowledge in isolation from its context.

Nicole Corbiere, an Anishinaabekwe master's student of Roy-Leveillée, advises researchers who want to partner with Indigenous groups to come prepared: "I think they also need to take the time to learn from other people and our research prior to going into communities."

For his part, Ernie Francis said he's noticed the shift in academic attention. But will such research effectively guide governments and policymakers? Francis believes it should: "Because at the end of the day, there's a lot of carbon that's going to come out of that permafrost."

To predict future effects of climate change, including increasing global temperature, scientists often rely on recorded global emissions of greenhouse gases. "We have these estimates of what Asia is responsible for, what North America is responsible for," Murphy said. "But these tend to be human-dominated emissions from transportation, industry—pollution, basically, or methane release from cattle production."

In other words, scientists are trying to solve a puzzle without all the pieces.

Though monitoring stations are installed across the Arctic, they are only in certain parts of the region and don't always collect information year-round. It becomes difficult, then, not just to collect data that provides the full picture, but also use it to compare emissions from permafrost to those released by industries, countries, and continents.

Such comparative data on carbon and other emissions from permafrost thaw is important, according to Arndt, because models based on national and continental carbon emissions are used to set global reduction goals in international agreements. "Leaving out emissions from permafrost thaw is like leaving emissions of Japan or the United States out of climate considerations," he wrote in an email to *Undark*.

Permafrost Pathways is attempting to rectify that gap by collecting data year-round, from which the team can tell how much carbon dioxide, methane, and energy in the form of heat and evaporation is exchanged between the atmosphere and the landscape. They also measure air temperature, radiation, precipitation, soil temperature, soil moisture, and snow depth, which help explain the exchanges of gas and energy in the environment.

Those monitoring stations, though, need to be maintained, and many researchers don't have the capacity to live near the stations all year. Consulting with the local community may help work around this challenge: They plan to hire Inuit in Pond Inlet and Resolute Bay, for example, to tell them how the equipment can be maintained throughout the year, and help with the maintenance, from cleaning sensors to making sure wires remain connected and functioning.

One of the monitoring stations is near the Churchill Northern Studies Centre, an independent field station in Manitoba that employs Indigenous technicians, so the network contracts them for maintenance work. Arndt said the team continues to seek out community members, whether to teach those who are interested about the research or pay them to help with data collection.

At a local co-op meeting, for instance, "we were sharing what our whole project is about and specifically what type of equipment we were looking to install, and why we're interested in working in and around their community with photos of examples," Arndt said. "We also wanted to introduce ourselves and speak one on one with community members. We really want to make sure people know who we are, why we're there, and form open relationships with the community."

Hard data is critical when it comes to climate research. But researchers are increasingly conducting qualitative work too, which relies on and documents Indigenous ways of knowing.

Inuit culture is intricately tied to the land: hunting polar bears and whales; traversing trails on their *qamutiik*, a traditional Inuit sled; and living in igloos, temporary shelters from the cold. As temperatures warm, Inuit's safety and livelihood are not only threatened, but also their customs.

Street is primarily involved in qualitative research to document such experiences (PermafrostNet comprises multiple arms of research, including predicting permafrost change in Northern communities and hazards associated with permafrost thaw such as slumps and coastal erosion.)

"Quantitative data is great in showing 'what' is happening, but I think the power of this project and community-based research is in the 'so what,'" Street said of her work in an email to *Undark*.

In order to ensure active participation and collaboration, Street has reached out to hunters and trapper associations, as well as renewable resource councils, to help craft the questions she would later ask community members. Indigenous communities have also provided logistical support, such as bringing her on location to conduct interviews, and connecting Street with potential translators, should interviewees request them, who spoke the local languages Inuvialuktun or Gwich'in.

These steps were important, she said, because her research, which is still ongoing, has found that no two communities or people are experiencing changing permafrost in exactly the same way: Some struggle more with flooding, others report drying waterways, making boat travel more difficult.

Street has also interviewed people who are taking their own steps to adapt, building their homes away from the water, for example, or changing the trails they use to access land for hunting and harvesting. The work feeds into one of the larger goals of PermafrostNet, which is to collaborate with and "assist northern communities to plan for and manage a changing permafrost environment by providing specific strategies to mitigate effects that are hazardous or debilitating to existing infrastructure," according to the organization's website.

Ernie Francis, the Gwich'in man who showed Street around Inuvik, says consulting with locals is especially important when it comes to mitigating the effects of a changing climate. He's seen how Indigenous know-how has been ignored, for example, in building infrastructure on unstable land: "They don't rely on local knowledge," he said of certain construction projects in his community. "You don't know the drainage of the water or areas where it's potentially better to build."

**Permafrost underlies anywhere from 15 to 25 percent of the Northern Hemisphere and stores up to an estimated 1.6 trillion metric tons of organic carbon in the region, twice the amount currently held in Earth's atmosphere.**

Street says her research has also been strengthened by community guides and environmental monitors who are part of local councils and committees and who act as a resource and liaison for their larger community.

Michael Cameron, an Inuk from Salluit, hasn't worked directly with Street, but has worked as a community guide for other researchers studying permafrost thaw in Nunavik. He says local wildlife monitors and guides in Indigenous communities, such as himself, are increasingly working with researchers because it ensures there is an exchange of knowledge and expertise. In order to get a license to do research in these areas, most teams also have their proposals evaluated by the local hunters and trappers committees, hamlet governments, as well as territorial impact review and environmental assessment boards.

Roy-Léveillée, the PermafrostNet researcher, has worked with Cameron, and she said that such relationships have strengthened her research on the ground—for example, she's adapted her questions based on concerns shared by the community, and has been connected to locals she may not have heard from otherwise.

But those kinds of collaborations are a more recent development, she said, with funding bodies "trying to reevaluate how they assess research contributions so that it becomes broader and includes more things—perhaps, hopefully—communication to communities."

Research about thawing permafrost and the Indigenous communities that live on it has grown over the years. Permafrost Pathways alone has received $41.2 million in funding to do this work, plus a $5 million grant from Google. In Canada, researchers with PermafrostNet were awarded $1.65 million to train the "next generation" of permafrost experts by the Natural Sciences and Engineering Research Council of Canada.

Yet some researchers say the process of conducting permafrost research that involves Indigenous communities, values traditional knowledge equally, and effectively guides national and international climate policy still has room to improve.

According to Roy-Léveillée, some aspects of academic funding make it difficult to put Indigenous communities at the center of conversations about permafrost thaw. For example, Roy-Léveillée said that in the past, the Northern Scientific Funding Program, which is funded by the Canadian federal government and has supported Roy-Léveillée's research, encouraged young researchers to report their results to the communities they worked with—but did not provide the monetary support for them to do so. (Misha Warbanski, acting director of science and technology at Polar Knowledge Canada, which administers the program, said that NSTP awards are intended to be "supplementary in nature" and that "engagement and communications plans are typically required elements of funding proposals.")

While Roy-Léveillée said this dynamic has changed in recent years—and that the program is "a great asset to northern research"—the past funding issues are an example of continuing limitations that exist within academic research: "Going back to talk to the community doesn't count and is not valued at the moment."

Another shift Roy-Léveillée says she is seeing in permafrost research is the focus on "circumpolar" studies, which often look at climate change across the Arctic rather than focusing on individual communities within it. Such research is now likelier to receive funding, she said, but often treads the line between ambitious and sensational: "There's a pressure to publish big, flashy circumpolar results."

The answer to these challenges may not always lie with engaging Indigenous groups, Roy-Léveillée said: Doing so may overwhelm smaller communities if researchers are adamant about working together. "I'm not into forcing people to interact with a community if they don't have that in them, because I think it can be harmful," she said. "Some people really want to do it and should be supported in doing it, and maybe the others should be left to do their thing and go home."

Corbiere, the Anishinaabekwe master's student, said she would like to see both traditional and Western knowledge coexist, with Indigenous knowledge providing historical context and data, and Western science focused on present and future predictions. "I don't think it's necessarily interweaving the Western knowledge and traditional knowledge," she said. "I think it was already there. It was our knowledge, and it was the knowledge of the keepers of traditional lands that has always been there. It's always been scientific."

With no single answer to connect Indigenous knowledge and permafrost research, Corbiere said the most important thing for both groups may be to keep working towards a trusting relationship.

"In my community, we have our Seven Grandfather teachings: It's wisdom, love, truth, respect, humility, bravery, and honesty," she said. "Bringing those teachings into my research I think is what has always helped me."

## Print Citations

**CMS**: Jamal, Meral. "Valuing Indigenous Knowledge in Permanent Research." In *The Reference Shelf: U.S. National Debate Topic 2025–2026: Exploration & Development in the Arctic,* edited by Micah L. Issitt, 49–54. Amenia, NY: Grey House Publishing, 2025.

**MLA**: Jamal, Meral. "Valuing Indigenous Knowledge in Permanent Research." *The Reference Shelf: U.S. National Debate Topic 2025–2026: Exploration & Development in the Arctic,* edited by Micah L. Issitt, Grey House Publishing, 2025, pp. 49–54.

**APA**: Jamal, M. (2025). Valuing indigenous knowledge in permanent research. In M. L. Issitt (Ed.), *The reference shelf: U.S. national debate topic 2025–2026: Exploration & development in the Arctic* (pp. 49–54). Grey House Publishing. (Original work published 2024)

# We Went to Greenland to Ask about a Trump Takeover

By Ben Schreckinger
*Politico*, January 10, 2025

NUUK, GREENLAND—At a recent Friday night *kaffemik*—a traditional, coffee-fueled gathering—Jørgen Boassen reclined at the end of a long table covered in bits of leftover whale skin and cake, wearing a T-shirt emblazoned with the words "American Badass" and the famous image of a bloodied Donald Trump raising his fist in the air.

Such brazen pro-Trump displays have made Boassen conspicuous—and famous—here in Greenland, an autonomous territory of the Kingdom of Denmark. But the 50-year-old bricklayer insists that as the island's inhabitants strive for full independence from their old colonial patrons, there is more support for the incoming American president here than meets the eye.

"Many want to use him to liberate us from Denmark," he said, raising his eyebrows suggestively.

That prospect, laughed off when the idea that Trump might "buy" Greenland from Denmark was first floated five years ago, has come roaring back.

In recent days, Trump's increasingly insistent posts about his desire to make Greenland part of the United States have set off an international furor. At a press conference on Tuesday, he threatened Denmark with tariffs over the island, refusing to rule out using military force to take it, as Danish Prime Minister Mette Frederiksen insisted that "Greenland belongs to the Greenlanders." Meanwhile, Denmark's King Frederik has updated the royal coat of arms to more prominently feature a polar bear that symbolizes the island's place in his realm.

Amid it all, Donald Trump Jr. traveled to Nuuk this week, where none other than Boassen showed him the sights on Tuesday.

There are real strategic reasons for the U.S. to seek closer relations with the island. Great power competition is heating up in the Arctic, with Russia and China increasingly active in the region. Greenland is located on NATO's northern flank and contains large deposits of rare earth minerals, essential ingredients in smartphones and car batteries.

That said, Trump's wildest ambitions face serious obstacles. The days when the U.S. government could outright purchase chunks of the Western hemisphere from a European power have passed. Even if Denmark were inclined to simply

From *Politico*, January 10 © 2025. Reprinted with permission. All rights reserved.

sell Greenland to the highest bidder, Greenlanders enjoy too much legal power over their island for that to happen.

The most recent opinion survey shows the Greenlandic public prefers several other potential international partners over the U.S., and, in public statements, the leaders of its government stress a desire for full independence, not annexation into another country.

At the same time, Greenland has been drifting out of Copenhagen's orbit for decades as its people have gradually gained greater rights of self-determination. But full sovereignty remains a tall order for the fewer than 60,000 inhabitants who would need to take over full responsibility for defending and developing an island three times the size of Texas.

All of this means, according to discussions with observers of Arctic geopolitics, that some sort of deal in which the U.S. and Greenland cement a special relationship, with the blessing, or at least acquiescence, of America's Danish allies, is not so far-fetched. Officials from Trump's first administration have already begun advocating for a treaty arrangement, called a free association agreement, with the island, including in a recent *Wall Street Journal* op-ed. (Such a deal would require Greenland to become independent from Denmark first.)

As for Boassen, he sees only opportunity for himself and his country. He is in regular touch with Tom Dans, a onetime Trump appointee to the U.S. Arctic Research Commission who runs a foreign policy nonprofit, American Daybreak, and has been advocating for closer ties between the U.S. and Greenland. Boassen is eager to pursue trade deals, and perhaps other arrangements, with the U.S.

But first he will have to persuade his fellow Greenlanders, many of whom remain apprehensive about the overtures from the superpower next door.

Another *kaffemik* guest Allen Henson, 37, offered a glimpse into just how much unease America's shifting foreign policy is causing here.

"I had a dream we got invaded," Henson confessed. "There were jet planes everywhere."

One way or another, change is coming to Nuuk.

The first signs of it greet visitors at the airport, which, thanks to Danish government backing, reopened with a modern international terminal a week before my arrival in early December. To reach Nuuk from Washington, I had to first fly over Greenland, then take a prop jet from Iceland's capital, Reykjavik. In June, United Airlines is set to begin seasonal direct flights from Newark to Nuuk, making travel between the two landmasses considerably more convenient, and the growth of American influence all but inevitable. Already, the internet has helped English gain a foothold here alongside the indigenous Greenlandic language and Danish.

But Greenland still remains enmeshed in the Kingdom of Denmark. While the island has gained increasing control over its internal affairs in recent decades, most of its government budget comes in the form of an annual subsidy from Copenhagen, which maintains control of Greenland's external affairs. Danish firms dominate much of the island's economy.

Whether or not full sovereignty is a realistic option for an isolated island of 56,000-odd souls, it is a popular goal, and anti-colonial sentiment runs strong here. In his New Year's address earlier this month, Prime Minister Múte Egede decried "the shackles of the colonial era" and renewed his calls to leave the Kingdom of Denmark.

Over lunch in downtown Nuuk, Rasmus Leander Nielsen, a professor at the University of Greenland, a bearded Dane who has lived here for the better part of a decade, said that Greenlanders have discussed transitioning to a free association arrangement with Denmark—in which Greenland would gain full sovereignty but maintain special economic, military and immigration privileges with Copenhagen.

But recent proposals from former Trump officials for free association with the U.S. have also been noticed here, he said. (Such proposals are modeled on U.S. arrangements with a handful of Pacific islands like Micronesia that grant the U.S. exclusive military access in exchange for economic benefits such as development grants and infrastructure funding.)

Nielsen recounted a recent phone call he had with a Greenlandic official, whom he declined to name for the record, in which the official floated a new idea: after independence, free association agreements with the U.S. *and* Denmark.

As Russia and China step up their presence in the Arctic, Greenlanders are increasingly acknowledging that American security guarantees, which already exist with Denmark and Greenland, through NATO, will be an indefinite fact of life in this corner of the world.

"Push come to shove," Nielsen said. "It's still the U.S. that is going to save our ass."

Boassen has his own vision: Embrace Trump, and make some sort of deal with the U.S. that will enrich Greenland while breaking it free of its dependence on Denmark.

On a whirlwind tour of Nuuk's eating and drinking establishments, he gave me the local lay of the land from a Trump-centric perspective. He was sure to point out who in town is a closet Trump supporter, and who, like a former mayor whose voice came on his car radio during a news segment, is a Kamala Harris supporter who no longer speaks to him.

At every stop in Nuuk, locals approached him to tease or congratulate him about the election results. At Unicorn, a restaurant in the old town, revelers at Christmas office parties drank and gorged on reindeer. Some wore costumes for the occasion, and upon our entrance, a man dressed as Trump dressed as a garbage collector made a beeline for Boassen, then asked for a photo.

Boassen, who has a beard and stocky build, is of mixed Inuit and Danish heritage, like many Greenlanders. Also like many Greenlanders, he nurses resentments toward the distant European power, which at times has imposed policies of forced cultural assimilation and birth control here. It does not help that as a child he had light skin and blonde hair and was beaten up for looking too Danish.

On a more practical level, he said that Greenlanders are chafing under inflation, and he points at the market power enjoyed by Danish businesses over many of the goods Greenlanders consume.

The Danes, he said, are more worried about losing Greenland than they let on.

"I think many Danish government members know me," he said. "They fear me. I think. I don't know."

He recounted an argument with a conservative Danish man who pointed to the treatment of Native Americans in the U.S. and warned him that America would forcibly relocate Greenlanders to Disko Island, off the west coast of Greenland north of here.

Boassen said he harbors no illusions that the United States and its incoming president are motivated by selfless love of Greenlanders, but he believes that they offer his people their best leverage against Denmark. As for the specifics: Boassen said the ultimate shape of a deal with the U.S. would be up to the people of Greenland. "Of course he cannot buy us," he said, "but we can be partners with [the] U.S."

Two centuries after President James Monroe declared the Western hemisphere off limits to European colonial powers, the world's largest island has remained a mostly overlooked exception.

At its closest point, the island is just 16 miles east of Canada, and scholars believe that the first humans to eke out an existence here were Indigenous Americans, with roots in northeast Asia, who arrived thousands of years ago. More is known about early European settlers. Famously, the Norse explorer Erik the Red sailed West from Iceland, named the island he encountered "Greenland" and established settlements there toward the end of the 10th century.

Those settlements eventually died out, but Europeans returned in the 18th century under Danish sponsorship. The Danes converted many of the Indigenous Inuit people they encountered to Christianity and made the island part of their kingdom.

U.S. interest in Greenland, meanwhile, dates back at least to the mid-19th century. Around the time that Secretary of State William Seward was acquiring Alaska from Russia, he explored a purchase of both Greenland and Iceland from Denmark, but did not see it through.

During World War II, the U.S. assumed de facto control of the world's largest island and established Thule Air Base in the far North. After the war, the Truman administration offered Denmark $100 million worth of gold for Greenland, but was rebuffed. The State Department retreated from the island in 1953, when it closed its consulate. The U.S. air base—which is now named Pituffik and hosts an early warning system for ballistic missiles—remained open, thanks to an agreement between the U.S. and Denmark.

Since then, the Greenlandic people—a group that is mostly Inuit, with significant Danish admixture—have won greater rights of self-determination from Copenhagen and charted a slow course toward independence.

One milestone came in 1979, when Greenland won home rule and established its own parliament. Another came in 2004, when it became a signatory to the World War II-era defense agreement between Denmark and the U.S. In 2009, Greenlanders assumed fuller control of the island's internal affairs, including its natural resources.

As Greenland has inched closer to independence, its melting glaciers have attracted the world's attention as a symbol of climate change.

It also started to attract strategic interest from the U.S. government.

As early as 2007, the State Department was eyeing a renewed push into Greenland, a diplomatic cable sent from the U.S. Embassy in Denmark to the CIA and other agencies, and later published by Wikileaks, reveals. The cable notes Greenland's drift toward independence as well as growing Chinese interest in the island's resources. It calls for the State Department to establish, after consultations with Denmark, a seasonal presence there.

"With Greenlandic independence glinting on the Horizon," the cable argues, "the U.S. has a unique opportunity to shape the circumstances in which an independent nation may emerge."

U.S. interest in Greenland remained the stuff of private, long-range government planning until news broke in August 2019 that Trump had become preoccupied with the idea of buying the island.

The idea reportedly sprang from a conversation with Estée Lauder heir Ronald Lauder, president of the World Jewish Congress, though a person close to Lauder said that notion of an outright purchase originated elsewhere. "Lauder never said to buy Greenland," said the person, who was granted anonymity to discuss a private conversation. Instead, the person said, Lauder merely told Trump it was "in our interest to engage more, have deeper ties."

Arkansas Republican Sen. Tom Cotton has said that he also urged Trump to purchase the island, and he reportedly floated the idea in a 2018 meeting with Denmark's then-ambassador to Washington, Lars Gert Lose.

Trump went public with his interest, likening it to a real estate deal and tweeting a mocked-up image of a Trump Tower skyscraper looming over a grouping of modest Greenlandic homes.

The notion was treated as a non-starter at best and a provocation at worst.

Some of the stumbling blocks to a U.S. purchase were practical: Denmark provides most of Greenland's government budget in the form of a $500 million annual subsidy, which allows the island to fund a Scandinavian-style welfare state. The U.S. would, presumably, have to take on that cost.

Others were more fundamental: Many Greenlanders were especially offended by the notion that the U.S. could buy their island because there is no land ownership in Greenland, where all land use is allowed by government permit.

Greenlandic officials settled on a stock response to inquiries: "Greenland is open for business, but we're not for sale."

The float also complicated U.S. relations with Denmark, which are marked by close security cooperation aimed at countering Russia, and increasingly, China, which has declared itself a "near-Arctic state." Earlier in Trump's term, when Beijing offered state-backed loans for three new airports in Greenland, the Pentagon had leaned on Denmark to offer financing in a successful bid to box the Chinese out of the projects.

But when Danish Prime Minister Frederiksen called Trump's purchase idea "absurd," the then-president retaliated by canceling an upcoming state visit to Copenhagen.

The idea faded from the headlines, but the Trump administration continued to pursue deeper direct relations with the Danish territory.

In June 2020, the State Department reopened a consulate here after a 70-year absence. Some in and around the administration continued to nurse more ambitious ideas: everything from increasing trade to adding it as the 51st state.

"I worked on Greenland until the final day of the last administration," said Dans, the Trump Arctic Research Commission appointee who is working with Boassen, in an interview. "All of the same reasons that existed then for increasing our partnership exist today."

The arrival of the Biden administration restored an Obama-era emphasis on multilateral cooperation and climate leadership to U.S. engagement in the Arctic.

But in the wake of Trump's victory, America's role in the world is once again poised to shift, and big ideas about Greenland have resurfaced.

In fact, the idea of increased U.S. involvement in Greenland has been percolating in policy and tech circles for months: A week before the election, Mike Solana, a venture capitalist at Peter Thiel's Founders Fund, hosted the latest installment of his "Hereticon," a gathering to discuss provocative ideas, at the luxe Faena Hotel in Miami. Nick Solheim, co-founder of American Moment, a group dedicated to staffing Trump's second term, gave a talk about acquiring Greenland, telling attendees that Trump had been dead serious about the idea. Solheim alluded to the Homestead Act, a 19th century law that incentivized the settlement of the American West, according to a person present, raising the prospect of Americans settling the island. Solheim declined to comment.

> **Great power competition is heating up in the Arctic, with Russia and China increasingly active in the region.**

Two days after the election, Georgia Republican Rep. Mike Collins posted on X an electoral map of the results with Greenland tacked onto it in red, along with the caption "Project 2029." (Some in Nuuk remarked to me that, if admitted to the union, their island would not be a red state.)

Meanwhile, an entrepreneur from California is pushing an even stranger vision: Dryden Brown, the founder of a startup that aims to create high-tech colonies around the world, drew attention with posts on X suggesting that American interests should use Greenland to practice making Mars habitable for humans. ("Nice fantasy," conservative Danish parliamentarian Rasmus Jarlov responded, "But forget it." In an email, Jarlov declined to comment to further on the matter, writing, "I will not discuss something that far-fetched in the media. I will leave that for teenagers on Twitter.")

Despite the renewed talk of colonies and buying, others in Trump's orbit are proposing arrangements that are more in line with recent precedent in international affairs, and therefore, perhaps, more realistic.

In the weeks before the election, Dans and Alexander Gray, who served as chief of staff to Trump's National Security Council, publicly proposed that the U.S. and Greenland enter into a Compact of Free Association. After the election, Gray doubled down on the idea in his Wall Street Journal op-ed.

"At this point it's less about the details," Gray said in an interview, "than there's a coalescing of views that a greater U.S. commitment to Greenland is a logical and strategically defensible approach."

The issue of the island's future is complicated by the delicate three-way dynamics at play between the U.S., Denmark and the Greenlandic people.

While Danes might not like the idea of the U.S. moving in on their kingdom's largest landmass, they have a close security relationship with the U.S., focused on countering Russia. As Trump's 2019 snub of Frederiksen—whose office declined to comment for this story—showed, protesting too loudly could damage it.

Then there's the Greenlandic people. The 2009 self-governance law passed in Denmark grants them decision-making power over their independence and outlines the process for a referendum, while stipulating that the terms of independence would be subject to negotiation with the Danish government. While the law says Denmark's parliament must consent to an independence agreement, it is not clear what would happen if Greenlanders voted for independence and Denmark's parliament voted against.

"A constitutional crisis run amok," Nielsen predicted.

Most Greenlanders—two-thirds in one 2019 poll—aspire for eventual independence, but the stately pace at which they have pursued it could frustrate the ambitions of a term-limited American president. In 2023, after four years of study, a parliamentary commission presented a draft constitution for an independent republic, but its implementation remains theoretical.

In a chance encounter at the Nuuk airport, Finance Minister Erik Jensen, who is also leader of the island's dominant party, Siumut, said he expects the next round of parliamentary commission recommendations to come around 2026.

As for independence itself, it is too soon to say when that might happen. "In our organization we don't discuss the year," he said. "But we would like to move faster."

Full independence would give Greenlanders the ability to control their own relations with the U.S. How close they would want those relations to be remains to be seen.

Fifty-nine percent of them would like to see more cooperation with the U.S., according to survey results published last month by Nielsen. But an even greater number, two-thirds, say they would like to see greater cooperation with Denmark and with the EU. Canada, Iceland and the Arctic Council, a multilateral organization, all rank even higher in the survey.

The stance of Greenlandic officials reflects those results. Many say they welcome more trade with the U.S., but they appear unenthusiastic about more ambitious American proposals.

"It depends of course on a more realistic approach than what we're seeing now," said Aaja Chemnitz, one of Greenland's two representatives to Denmark's parliament, who predicted that Copenhagen would remain Greenland's primary international partner.

"I haven't seen many results from U.S. engagement with Greenland," she said.

That may be in part because of the current government's reticence towards American overtures, even when they come from relatively high-ranking officials.

In late November, when Undersecretary of State Jose Fernandez traveled to Nuuk to discuss development of the island's mineral resources, he was unable to secure a meeting with Prime Minister Egede.

"Due to the business in the meetings at the parliament, the prime minister was not able to find time," said his spokesperson, Julia Ezekiassen, who declined to make Egede available for an interview.

A State Department spokesperson touted Fernandez's meetings with other senior officials and private sector representatives.

While a prime minister snubbing an undersecretary of state might be expected in many places, the size of the two polities involved makes it notable: The State Department has more employees than Greenland has citizens.

Local feelings about far-off power centers figure prominently in the local politics and culture: The government titled its most recent foreign policy road map "Nothing about us without us."

During my visit, Nuuk's modest art museum was restaging an old exhibit that imagines Greenland as a belligerent expansionist power, complete with agitprop extolling the virtues of its attack kayaks. A map hanging on the wall imagined a Denmark colonized by Greenland, with its regions renamed for famous islanders. The original intent of the project was satirical, one museum staffer explained, but she said that these days, not everyone takes it that way.

At a Friday night concert at a community center, Nuuk's young and old gathered to dance. A local musician, Tupaarnaq Ingemann Mathiassen, sang a song in Greenlandic lamenting the removal of Indigenous people to make way for the U.S. air base during World War II: "The kind people with no evil inside ... will not see their land again."

Boassen appeared unbothered by the song. He is more interested in what the U.S. and its incoming president could do for Greenland's future.

Boassen argues the incoming president could be a liberating figure around the world. He cited the first Trump administration's launching of a task force to address the problem of missing Native women and said that such actions set a tone that strengthens the hand of Indigenous people elsewhere.

Though he conceded that Trump is unpopular here, he blamed the island's reliance on Danish news outlets, which he accused of tilting toward Democrats.

But between Trump's victory and the growing influence of social media here, the political status quo here may be primed for disruption. Boassen certainly detects an opening.

He said that while his Facebook posts in Danish are often flagged for moderation, he can post more freely in Greenlandic. He showed me an inbox stuffed with private messages and said that people from all over the island's small, scattered settlements agree with his pro-Trump writings.

Boassen also showed me the protest videos proliferating on TikTok after the Danish government determined a woman of Greenlandic extraction was not competent to raise her newborn child and separated them.

He views it as another sign that Greenlanders are more ready for change than the island's current leaders acknowledge—and said he is ready to press the advantage.

This past fall, he traveled to the U.S. and knocked on doors in Pennsylvania with Dans. Then he attended Trump's election night rally at the Palm Beach Convention Center, snagging a photo with Donald Trump Jr.

On Tuesday, when Trump Jr. arrived in Nuuk, Boassen was waiting to greet him inside the airport terminal, sporting a white anorak, a traditional Greenlandic jacket. Boassen said that he got a call two days prior from Greenland's diplomatic office in Washington, relaying a request for his contact information.

Boassen said he took Trump Jr. to see a handful of sights around Nuuk and then to a gathering of local Trump supporters at the steakhouse atop the Hotel Hans Egede, the city's premier lodging for international travelers.

Boassen said that Trump Jr. quizzed him about Greenland's culture and climate, and about local attitudes toward the U.S. and Denmark. "Not so much about politics," he said. Andy Surabian, a spokesperson for Trump Jr., declined to comment.

There will be plenty of time for politics. Boassen and Dans are working to put together a Greenlandic delegation to Washington for the inauguration. To promote exports, Boassen wants to present Trump with a sealskin coat.

His plans do not end there. By April, Greenland is due to hold elections. The vote will test Boassen's conviction that his homeland is ripe for change.

He intends to run for a seat in parliament on a pro-Trump, pro-America platform. He has the pitch down cold. The only question is whether Greenlanders are ready to buy it.

## Print Citations

**CMS**: Schreckinger, Ben. "We Went to Greenland to Ask about a Trump Takeover." In *The Reference Shelf: U.S. National Debate Topic 2025–2026: Exploration & Development in the Arctic*, edited by Micah L. Issitt, 55–64. Amenia, NY: Grey House Publishing, 2025.

**MLA**: Schreckinger, Ben. "We Went to Greenland to Ask about a Trump Takeover." *The Reference Shelf: U.S. National Debate Topic 2025–2026: Exploration & Development in the Arctic,* edited by Micah L. Issitt, Grey House Publishing, 2025, pp. 55–64.

**APA**: Schreckinger, B. (2025). We went to Greenland to ask about a Trump takeover. In M. L. Issitt (Ed.), *The reference shelf: U.S. national debate topic 2025–2026: Exploration & development in the Arctic* (pp. 55–64). Grey House Publishing. (Original work published 2025)

# No. 21 | The Arctic Council and the Crucial Partnership Between Indigenous Peoples and States in the Arctic

By Edward Alexander and Evan T. Bloom
*Wilson Center*, July 27, 2023

When the Arctic Council is functioning normally—something that we have not seen since many of its activities were "paused" as a result of the Ukraine conflict—it provides an important and indeed unique forum for cooperation between Arctic Indigenous Peoples and the eight states that are members of the Council. As an Indigenous group leader and a former diplomat, respectively, the two authors have witnessed first-hand the value of that cooperation as well as its limitations. We recognize the difficulties posed to the Indigenous Peoples when the Council doesn't function, and the promise it brings when it does.

In this article, we offer some reflections on the relationship between the Council and its Permanent Participants, what it has achieved in the past, and where it might lead in the future.

## Establishment of the Arctic Council

The Arctic Council was established in 1996 based on a proposal by Canada to transform the existing Arctic Environmental Protection Strategy into a high level forum focused on environmental protection and sustainable development. Canada's initiative was conceived of and promoted by its Inuit leaders, in particular Mary Simon (who is now Canada's Governor General). The idea was to bring together the eight states with territory above the Arctic Circle (Canada, Denmark, Finland, Iceland, Norway, Russian Federation, Sweden, the United States) with major Indigenous organizations. Within the Council, these organizations, referred to as the Permanent Participants, and states would sit as equals, although formal decision-making resides with the states. In 1996, the original Permanent Participants were the Inuit Circumpolar Conference (now the Inuit Circumpolar Council), the Saami Council, and the Association of Indigenous Minorities of the North, Siberia and Far East of the Russian Federation (now known as the Russian Association of Indigenous Peoples of the North – RAIPON). They have been joined by the Aleut International Association, the Arctic Athabaskan Council, and the Gwich'in Council International.

---

From *Wilson Center*, July 27 © 2023. Reprinted with permission. All rights reserved.

The two-year chairmanship of the Council rotates among the states, with a meeting of ministers held at the end of the chairmanship. The work of the Council is led by the Senior Arctic Officials from each state and the senior representatives of the Permanent Participants, and its main activities take place in six working groups as well as expert groups and ad hoc bodies established to tackle particular issues.

## A Historic Success

The relationship established via the Arctic Council between states and indigenous representatives is unique in diplomacy. While it is usual in international organizations to create a category of observers, including indigenous groups and non-governmental organizations, who can come to meetings to advise states, comment on the proceedings and watch what occurs, the Permanent Participants are not observers. They are an integral part of the institution. That states and the Permanent Participants sit together as equals in a multilateral organization is not something found elsewhere.

The Permanent Participants ("PP's" in Arctic Council parlance) ensure that the views of the Arctic Indigenous are a vital part of the work of the Council, providing a more direct and effective perspective than if the states relied on consultations with their Indigenous citizens for information and guidance. The PP's also represent in most cases a cross-section of nationalities, which provides cross-border perspectives. The Aleut International Association (AIA), for example, consists of membership in the United States of America and Russia. The Inuit Circumpolar Conference (ICC) consists of membership in Canada, the Kingdom of Denmark, the United States of America, as well as Russia. In the current political climate it's important that we collectively understand there is more than one avenue of diplomacy.

Not everyone appreciates why Indigenous participation is important in a body like the Arctic Council. One reason is that the Arctic Indigenous over the centuries have built up knowledge about conditions in the Arctic and how to thrive there, often in naturally sustainable hunting and fishing economies, and always in deep connection with nature. Their ancient knowledge is an essential element for their continued survival in the region, and their contemporary understandings about the changes they've experienced in the Arctic are needed for sound public policies for the region. After all the Arctic is their homeland, they have human rights and are the essential stakeholders in the region, and often, though not always, comprise the vast majority of the inhabitants of the area. Thus, their participation provides substantial value that the governments of the states (national and subnational) don't provide.

The diversity of the Indigenous representation within the Council is also important. Although many in, and outside, the Arctic think primarily of maritime matters given the importance of the Arctic Ocean, many Indigenous live far away from coastal areas. In these inland communities, their lives focus on non-maritime concerns, including these days the impacts on infrastructure caused by thawing permafrost, policies with implications for managing wildfires, the health of rivers and streams, and the fish that are found there. The Gwich'in live in-land and provide a perspective that may be different from coastal Indigenous brethren. The Arctic isn't just about the sea ice or shipping lanes. Indeed, much of the story of the Arctic can't be told as a maritime tale; for example, the setting of the Arctic and subarctic includes the largest forest on planet Earth, the boreal forest.

## Indigenous-led Arctic Council Projects, Participation and Impact

Gwich'in Council International has two projects centered on impacts to the boreal forest from wildland fire within the Arctic Council currently: one in the Conservation of Arctic Flora and Fauna Working Group (CAFF) and the other in the Emergency Preparedness Prevention and Response (EPPR) Working Group. Both projects were spurred by the devastating fires Gwich'in communities, many north of the Arctic Circle, have already endured. Some 64 percent of the Yukon Flats National Wildlife Refuge, home to the Gwich'in, have burned since 1960. This is the third largest wildlife refuge in the United States and is roughly five times the size of the State of Delaware. This area was established as a refuge for its unique status as a vast Arctic wetland, with much of it covering permafrost soils (permanently frozen soil rich in sequestered greenhouse gases), much of which has been impacted by the wildfires in the Arctic. Some four Delawares worth of land have burned within Gwich'in homelands alone.

The Gwich'in projects at the Arctic Council hope to draw attention and science to the Arctic to better understand the changes to wildfire in the Arctic and its impacts that they have already seen, as well as to draw international cooperation and resources to the issue. Northern latitude forests and Arctic tundra have removed tremendous amounts of $CO_2$ from the atmosphere over time, and that carbon has been stored (sequestered) in plants and especially soils. For example, circum-boreal forests contain more carbon than any other forest types on Earth (1,095 billion tons of carbon), and permafrost soils (found in both northern boreal and Arctic regions) store up to of 1,580 billion tons of carbon. To put these amounts into perspective, the carbon stored in permafrost is twice what is currently in the atmosphere (860 billion tons of carbon), meaning that permafrost thawing or burning from wildfires could release that stored carbon back into the atmosphere as the greenhouse gas $CO_2$, accelerating global warming.

The situation is terrifying and clear: the largest terrestrial stores of carbon on the planet are flammable, and we have increasingly seen large mega fires in the north burning millions of acres, and releasing vast and unquantified amounts of greenhouse gases. In addition, permafrost regions have historically served as a natural insulator keeping the sequestered greenhouse gases in soils contained, but, as the Arctic warms the vast Yukon Flats and other areas in the far north have shown this is less true now than before. As Gwich'in Council International have publicly stated their concern, the region has warmed three times as fast as the rest of the globe, and there is an acute need to better understand how much of the carbon stored in our northern ecosystems will be released to the atmosphere, and how fast will this occur given rapid Arctic change.

The Arctic Council is critically important for the space it creates for Indigenous perspectives, and leadership on issues like wildfire response in the Arctic. GCI's projects create international space for cooperation and awareness on critical issues like the climate driving impacts of wildland fire in the Arctic and subarctic regions. The CAFF project, Arctic Fire, hopes to expand our understanding of wildland fire ecology in the Arctic, exploring Indigenous knowledge on the topic, and creating space for experts around the circumpolar north to discuss these timely issues. The GCI project at EPPR, Circumpolar Fire, evaluates the legal frameworks of current wildfire cooperation between Arctic States and has proposed that new areas of cooperation and agreements may be warranted in the current climate of Arctic wildland fire. Naturally, these are only a couple of the examples of the many important projects Arctic Indigenous Peoples are engaged in at the Arctic Council.

The Indigenous groups also in many cases represent peoples who live in different countries. The Gwich'in communities are located in both Canada and in the United States. Thus, the perspective of Indigenous groups can be broader than found in a single nation, and can point to new areas of needed bilateral cooperation between neighboring states that can sometimes be forgotten in southern capitals. For example, the Indigenous groups have led efforts to focus on the importance of removing obstacles to travel across borders where families of particular tribes or groups live in different countries. Only recently, President Biden and Premier Trudeau recognized the need to improve conditions for visa-free travel between the United States and Canada to aid contacts between these families, and it was Indigenous groups who have pressed for this over many years. The complex nature of the Jay Treaty and Canadian immigration policy as they affect Indigenous Peoples are bound by a particular reality: these are one people, one language, families, who have lived and governed their own areas for tens of thousands of years, who have been separated by relatively recently established borders, and they have the right to continue their relationships.

The Indigenous are close to the land and nature and have a direct understanding of what is going on in the Arctic, especially environmental change which is at the heart of the issues the Council cares about. Whether it is intricate changes to sea ice, plant, animal, insect life, or ecosystem-wide rapid changes like wildfire, the Indigenous peoples are vested stakeholders in the Arctic who are keen and longtime observers of the region, and their knowledge is integral in getting things right for the Arctic.

The history of the Council has demonstrated its value as a forum that fosters communication between Indigenous groups and the central governments of the states. For various reasons, relationships between the Indigenous and ministries of governments, as well as domestic political processes, can be difficult. At times, the Permanent Participants can reach capitals through statements in the Council when other channels are not successful. On the margins of Arctic Council ministerial meetings, U.S. Secretaries of State have taken the time to have separate meetings with U.S. Indigenous leaders. In that respect, it appeared that groups representing Indigenous peoples living in the United States had more successful access to the U.S. Secretary of State than when they reach out to senior Alaskan state government officials.

The Arctic Council deliberately does not cover military issues. Because of the Ukraine conflict and resulting rise in tensions, and the expansion of NATO to include Finland, and soon Sweden, there is increasing attention given to military security issues in the Arctic. Those military security issues often decrease the attention to the core security questions of importance to Indigenous communities, such as food security, health, energy and jobs, as well as the changing environment. Focusing on the Indigenous interests helps bring the attention of states back to what is of greatest importance to the people who live in the Arctic. Further, the Indigenous people can help to draw the focus of the Arctic onto issues critical to the global community, like wildland fire in the Arctic or the shifting qualitative realities around permafrost melt and soil destabilization, that can sometimes go unseen in southern areas, or be overlooked for regional security concerns. They understand the value of their home and its qualities, and sometimes need to remind others that changes regarding it aren't always an opportunity for others to exploit.

> **There are many reasons to preserve the Arctic Council as an institution despite the tensions caused by the Ukraine conflict.**

## How Could Relations in the Council Improve?

The Permanent Participants will always be less-well funded than the officials and experts who governments send to Council activities, but more can be done to ensure that funding for their participation is adequate and assured. Many Indigenous delegates are in effect volunteers, not paid bureaucrats like the government officials. Some have funding to travel only to a few key meetings, like those of the ministers, Senior Arctic Officials or the Sustainable Development Working Group. Funding is needed not only for individual groups, but also for the Indigenous Peoples Secretariat. Currently only two member states fund the IPS, and state contributions to Indigenous Peoples organizations often don't reflect the value that they bring to the table.

The member state governments can work harder to ensure active communication. Many Indigenous People's Organizations felt that the states did not adequately consult with them prior to the Council's pause as a result of the Ukraine conflict. There was concern in the wider Arctic community about what the pause would do to Arctic Council cooperation, long valued and continued despite global geopolitics, and concern among some Indigenous People's organizations about increased military tensions in the Arctic at a time when scientific cooperation in the region is critically important to the globe.

There are many reasons to preserve the Arctic Council as an institution despite the tensions caused by the Ukraine conflict. If the eight Arctic States cannot find an accommodation that allows them to secure the existence of the Council until the Ukraine conflict has ended, Indigenous interests will be significantly harmed. Avoiding that harm does not mean that the states other than Russia should make concessions to Russia or accept the brutal treatment of Ukraine and its people, but there needs to be a recognition that if the Council can't continue, rebuilding it at some later time will take much time and effort, if it can be rebuilt.

Mary Simon and Canada showed great leadership in combining a country's interests with those of Indigenous Peoples in proposing the Arctic Council. The United States can learn from this example by not only accepting the importance of Indigenous representation and Indigenous Knowledge in international relations and agreements, but by becoming advocates for these in fora outside the United States as well.

## Print Citations

**CMS**: Alexander, Edward, and Evan T. Bloom. "No. 21 The Arctic Council and the Crucial Partnership Between Indigenous Peoples and States in the Arctic." In *The Reference Shelf: U.S. National Debate Topic 2025–2026: Exploration & Development in the Arctic*, edited by Micah L. Issitt, 65–70. Amenia, NY: Grey House Publishing, 2025.

**MLA**: Alexander, Edward, and Evan T. Bloom. "No. 21 The Arctic Council and the Crucial Partnership Between Indigenous Peoples and States in the Arctic." *The Reference Shelf: U.S. National Debate Topic 2025–2026: Exploration & Development in the Arctic,* edited by Micah L. Issitt, Grey House Publishing, 2025, pp. 65–70.

**APA**: Alexander, E., and E. T. Bloom. (2025). No. 21 The Arctic Council and the crucial partnership between indigenous peoples and states in the Arctic. In M. L. Issitt (Ed.), *The reference shelf: U.S. national debate topic 2025–2026: Exploration & development in the Arctic* (pp. 65–70). Grey House Publishing. (Original work published 2024)

# Understanding the Arctic Through Indigenous "Perspectives"

By Marta Asenjo Fernández
*REVOLVE*, July 9, 2024

There are many Arctics. This diverse region stretches across the northernmost part of the earth featuring a distinctive ecosystem and a variety of unique cultures. This interview with Marta Asenjo Fernández from *Revolve*, discusses the integration of local communities into scientific research, the current state of cooperation between researchers and these communities, and how international policies and external actors can and should protect the rights and traditions of Arctic indigenous Peoples.

What is the Arctic, and how does our perception—especially from Western countries—differ from that of local Arctic communities? How might these differing perspectives influence responses to climate change and collaboration with the scientific community?

Mia: The Arctic refers to the region around the Earth's North Pole, including the Arctic Ocean and adjacent landmasses. It is characterized by its extreme cold temperatures, unique ecosystems, the presence of sea ice, permafrost, and tundra as well as indigenous peoples and their livelihoods. From the perspective of Western countries, the Arctic has often been viewed through natural resource exploitation, strategic importance in terms of geopolitical changes, and scientific actions as the area warms at least four times faster compared to global average.

However, the perspectives of local Arctic communities differ significantly even among those communities living in the western culture (for example, the Nordic countries). For indigenous peoples such as the Inuit, Saami, and others, the Arctic is not just a remote wilderness but their home, deeply intertwined with their cultural identity, traditional knowledge, and ways of life. Their relationship with the environment is more holistic, enriched by age-old traditions and cultures.

These differing views can influence perceptions of climate change and collaboration with the scientific community, and they can also influence the ways the indigenous scientists see research interests, priorities, and goals. The indigenous communities represent also rightsholders, stakeholders, and scientists—not only traditional ways of life and traditional knowledge systems.

From *REVOLVE*, July 9 © 2024. Reprinted with permission. All rights reserved. https://revolve.media/interviews/understanding-the-arctic-through-indigenous-perspectives.

The ongoing green transformation (in the Nordic Arctic, such as Finland and Sweden) has already caused conflicts over land use, resource extraction, and conservation efforts, not only among indigenous peoples but local people in the Arctic in general. Coproduction of knowledge by combining different knowledge systems—scientific and traditional.

Arctic knowledge, and transdisciplinary research is needed to be able to tackle the challenges of Arctic transformation and related questions of justice. Building trust and fostering genuine collaboration with local communities is essential for effective research and policymaking in the Arctic addressing the complex challenges, such as those driven by a changing climate.

Considering the historical context of socio-economic inequalities faced by Arctic indigenous Peoples, how might the increased presence of scientists and researchers impact these communities? Can you provide insights into potential strategies or approaches that could help address or mitigate any negative consequences, ensuring that scientific activities contribute positively to the well-being of local Arctic communities?

Mia: The increased presence of scientists and researchers in the Arctic can, for instance, affect local communities' time resources and cause "fatigue" to participate in these research actions. However, it can also help raise local voices via knowledge exchange—and by doing so affect planning and decision-making, as well as create job opportunities and economic benefits through employment, contracting, and collaboration. Moreover, if not carefully planned with local communities, and without field work training for researchers, some scientific activities can contribute to environmental degradation or cause disturbances to wildlife or traditional actions—such as reindeer roundups. Researchers must be familiar with the areas, local cultures, and follow ethical guidelines. Collaboration, respect, and equitable partnerships are necessary to avoid negative consequences.

How is the scientific community currently integrating local Arctic communities into their activities?

Mia: The scientific community is increasingly recognizing the importance of integrating local Arctic communities into their activities to ensure research relevance, ethical conduct, and positive impacts on the well-being of these communities.

Many scientific projects in the Arctic such as REBOUND now employ a community-based research approach, where scientists collaborate closely with local communities for the lifetime of the project. This ensures that research questions, methodologies, and outcomes are relevant and beneficial to the community and that the research results have societal and policy impact as scientists are increasingly engaging with policymakers.

Scientists are increasingly using participatory research methods that involve local communities in data collection, analysis, and interpretation and involve preferences of local communities regarding research priorities and goals. This not only enhances the quality and validity of research findings, fosters mutual trust, respect, and ownership, but it also empowers community members to contribute their knowledge and perspectives to the scientific process.

Without inclusive scientific approaches it is not possible to produce culturally relevant, socially responsible, and sustainable research outcomes that benefit both the scientific community and the communities living in the Arctic and implement just research practices and policies for the changing Arctic.

Henry: The track record is mixed. Overall, I would say that there has been some progress on paying closer attention to indigenous knowledge. Our [*researchers*] methods are improving, more is being done, and more attention is being given to issues like data sovereignty (that is, who controls the information, rather than just letting scientists do what they would like with what they learn).

One of the challenges is that it takes time and effort to work in this way, so there is often an incentive to take shortcuts or avoid working with Indigenous communities. For those who take the time and effort, however, there are many professional and personal rewards from working closely with others, learning from them, and discovering things we would never otherwise have thought of.

For Arctic communities, these collaborations can be a chance for their voices to be heard, for their ideas to be shared, so that the story being told is not one-sided.

There is still, however, much progress to be made in involving Indigenous knowledge holders in the steps beyond research. The information is often used to make decisions around issues that are intertwined with local people's lives, such as land and resource management, and those choices remain largely in the hands of people who are not from the Arctic and who have limited understanding of indigenous knowledge and ways of life.

Studies that engage with indigenous knowledge are good, but not an end in themselves. More needs to be done to respect and enhance sovereignty and self-determination for Arctic peoples.

How does the scientific community approach the integration of local indigenous communities in their research in the Arctic, given the historical background of marginalization, forced sedentarization, and land dispossession?

Henry: The track-record of scientists working with Arctic peoples is variable. Some scientists have historically worked very effectively and respectfully with local people. Unfortunately, other scientists have not been so respectful, and that continues all too often. It is essential to treat everyone with dignity and respect, which includes educating ourselves about Arctic peoples' history and ways of life to avoid assumptions and don't ignore the legacy of the past.

We may not be able to make up for all that has happened, but we can determine how we will interact with people from now onwards. Our choices are important and can help make small but meaningful differences over time.

In what ways can the scientific community effectively integrate indigenous knowledge into their research and activities? What ethical considerations and dynamics should be addressed to ensure equitable partnerships and respect for Indigenous perspectives?

Mia: Scientific research and activities require a collaborative and respectful approach that recognizes the value and validity of indigenous and local perspectives and involves indigenous scientists and indigenous stakeholders and considers traditional / indigenous / local / practitioners' knowledge. Active two-way communication is essential to include their perspectives, concerns, and priorities, as well as to build trust and mutual understanding while developing shared knowledge and goals.

> **For indigenous peoples such as the Inuit, Saami, and others, the Arctic is not just a remote wilderness but their home, deeply intertwined with their cultural identity, traditional knowledge, and ways of life.**

This includes acknowledging the expertise of indigenous elders and knowledge holders and incorporating their insights into scientific research and decision-making processes, as well as an awareness of cultural protocols and traditions when conducting research in Indigenous communities.

Henry: Indigenous knowledge is a valuable source of information and understanding, and there are many ways for scientists to engage with indigenous knowledge and (especially) with the holders of indigenous knowledge. The central focus should be establishing a mutually respectful and mutually beneficial relationship. A good basis for this is the idea of free, prior, and informed consent, to make sure that everyone is willingly participating and has the opportunity of withdrawing at any time.

How can scientists actively engage with Arctic Indigenous communities to harness their traditional knowledge effectively? Furthermore, how can this collaboration contribute to preserving indigenous cultures and fostering a sustainable relationship between these communities and the environment, particularly considering changing climate conditions?

Henry: Building relationships is essential to creating ethical partnerships. This can go from a short discussion to a major, multi-year project, or even a long-term program or institution. Scientists need to be open to learning about Indigenous communities and their ways of life and understandings of the world, which can be very different from scientific understanding.

It can be easy to dismiss the things that do not neatly fit within the scientific paradigm but, in my opinion, that is not respectful. Arctic Indigenous peoples have inhabited the region for millennia, and so it is better to start off with the idea that their knowledge has been proven over time, and that those of us from outside the Arctic have a lot to learn from them.

As for helping sustain Indigenous cultures, one insidious effect of colonization is that those who have been colonized (such as the Arctic Peoples) devalue their own knowledge. It would be positive for studies that respect indigenous knowledge to help increase awareness of how important and valuable the information compiled from these communities is.

How do you think the rights of Indigenous Peoples, as outlined in United Nations Declaration on the Rights of Indigenous Peoples (UNDRIP), can be better prioritized in adaptation policy development and support in the Arctic?

Mia: Prioritizing the rights of Indigenous Peoples, as outlined in the United Nations Declaration on the Rights of Indigenous Peoples (UNDRIP), in adaptation policy development and support in the Arctic requires respect of their traditional knowledge and cultural practices.

Their rights, interests, and perspectives should be incorporated into decision-making processes and give space for self-governance, for instance in matters of safeguarding sacred sites, traditional hunting and fishing grounds, and cultural practices that are integral to their indigenous identity and well-being for adaptation capacity building, shared decision-making processes based on principles of equality and reciprocity.

Given the growing political and economic interest in the Arctic from outside the region, how can Indigenous and local communities ensure their voices are heard and their rights are protected in decision-making processes?

Mia: Participatory processes need to actively involve Indigenous and local communities in all stages of planning, policymaking, and implementation. This includes ensuring meaningful consultation, collaboration, and consent in matters that affect their lands, resources, and well-being.

However, this requires national legislation, international agreements, and legal frameworks that support participatory decision-making and engagement.

In terms of research and data collection, community-based monitoring and research initiatives to document and address the impacts of environmental change, resource development, and other factors affecting Indigenous lands and livelihoods can greatly help. Media and other partnerships can also strengthen advocacy efforts and increase leverage in decision-making processes.

How can international cooperation and agreements help in protecting the rights of Indigenous Peoples and preserving the cultural and ecological integrity of the Arctic region?

Mia: International agreements can serve as platforms for recognizing and upholding the rights of Indigenous Peoples in the Arctic, including their rights to self-determination, land, and natural resources.

However, although agreements such as the UNDRIP provide a framework for governments and stakeholders to respect and protect these rights, it is not always self-evident that all indigenous peoples have the same rights even within the indigenous communities (between certain indigenous groups) because some indigenous groups are not officially recognized as such (for example, forest Sami in Finland).

Therefore, careful considerations should be made about how those rights are determined and by whom.

In terms of environmental conservation, international agreements can promote the conservation and sustainable management of Arctic ecosystems and biodiversity, but they can also violate some cultural and subsistence practices.

Certainly, international cooperation is essential for addressing the impacts of climate change and building resilience to climate change in the Arctic region. International agreements can also support efforts to preserve and promote indigenous languages, for instance, and help maintain cultural heritage through the United Nations Educational, Scientific and Cultural Organization (UNESCO).

The Arctic Council also provides a forum for Arctic states, indigenous organizations, and other stakeholders to collaborate for addressing common challenges and promoting sustainable development in the region. These agreements can facilitate dialogue, cooperation, and shared views on issues and policies affecting Indigenous Peoples and the Arctic's ecosystem.

## Print Citations

**CMS**: Fernández, Marta Asenjo. "Understanding the Arctic Through Indigenous 'Perspective.'" In *The Reference Shelf: U.S. National Debate Topic 2025–2026: Exploration & Development in the Arctic,* edited by Micah L. Issitt, 71–76. Amenia, NY: Grey House Publishing, 2025.

**MLA**: Fernández, Marta Asenjo. "Understanding the Arctic Through Indigenous 'Perspective.'" *The Reference Shelf: U.S. National Debate Topic 2025–2026: Exploration & Development in the Arctic,* edited by Micah L. Issitt, Grey House Publishing, 2025, pp. 71–76.

**APA**: Fernández, M. A. (2025). Understanding the Arctic through indigenous "perspective." In M. L. Issitt (Ed.), *The reference shelf: U.S. national debate topic 2025–2026: Exploration & development in the Arctic* (pp. 71–76). Grey House Publishing. (Original work published 2024)

# 3
# Sustainable Economic Development

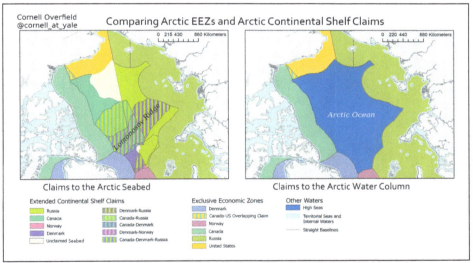

Claims to the Arctic Seabed and the Arctic Water Column. Illustration by Alphaomegapyat, 2021, CC BY-SA 4.0, via Wikimedia.

# The Search for Arctic Resources

The European colonization of the Arctic, and the exploration of this territory, was driven by the pursuit of profit and power. It was the Vikings who first visited the Arctic Circle in the ninth century, culminating in Erik the Red's colonization of southern Greenland, but for a long time thereafter, no Europeans visited the Arctic. It wasn't until the twelfth century that Russian explorers began moving through what is now Siberia, ultimately making much of the Arctic Russian territory by the seventeenth century.

This early exploration and colonization, by Russia and the Scandinavian kingdoms opened up the first trade in the Arctic, focusing on biological resources. Seal flesh and skins were shipped along trade routes to Europe and it became fashionable, for a time, to deliver polar bears, alive or dead, to the ruling heads of Europe's houses. There was a live polar bear kept in the menagerie in the Tower of London under the reign of Henry III.[1] But these novelty items and rare commodities didn't drive much commercial traffic to the Arctic, because the journey was too long and dangerous to be economically viable for most European traders.

There was one thing that kept bringing explorers back to these dangerous environments, despite the risks, the possibility of discovering the long sought "Northwest Passage," an imagined route through the continent connecting the Atlantic and Pacific oceans. This search was about power and profit as well, essentially constituting a new trade route that could be dominated by the nation that discovered the safest passage. Ultimately, a good route through wasn't really discovered until the 1800s, but the search for this passage brought many explorers to the Arctic including Christopher Columbus and others whose names have survived in the historic record. For more than four centuries, British, Norwegian, Russian, Danish, and Dutch explorers sailed to the Arctic in search for this route, with many ships and crews lost in the ice. Though the effort was ultimately a failure, these explorers "discovered" many new Arctic territories, and new trading companies were established to harvest resources directly from these islands and northern territories. The Hudson's Bay Company and the Royal Greenland Trading Company survived from the 1600s to the modern era, though their business practices have changed. At the time, walrus hunting and the harvesting of whale oil were the driving businesses. By the 1800s, hunters had driven these species to near extinction, and the trade had collapsed.

One of those English explorers, Martin Frobisher, claimed that he discovered gold on Baffin Island in 1576. This began a new wave of explorers heading to the Arctic itself in search of treasure, rather than seeking a route through the ice, who were looking for treasure. Baffin Island's "gold" turned out to be pyrite, which was worth very little then and less today, but explorers eventually found the kinds of resources they were hoping for, and it drove colonization and ultimately the vast exploitation of the Arctic realm. It was this leg of the search for Arctic treasure that resulted in the establishment of English control in Canada.

Over the years that followed, other forces drove exploration of the Arctic. First, there was the exploratory scientific revolution of the 1800s, which saw intellectual exploration of the Arctic realm in hopes of better understanding Earth and the Earth sciences. Then, during World War I and II, and especially in the years that followed the Second World War and became the Cold War, exploration and colonization of the Arctic was driven by military goals. This continues today, with the expansion of Russian militarization in the Arctic and the formation of new nuclear weapons sites. Military posturing and security claims are part of what has driven Donald Trump and allies to call for the US "purchase," which is not possible under international law, of the sovereign nation of Greenland.

In the 2020s, both military motives and economic advantages are used to justify European and North American interests in the Arctic, but the economic resources remain the primary draw for those who see an advantage in the Arctic, rather than environments worthy of protection and conservation. This is driven, first and foremost, by the reserves of fossil fuels located in the Arctic realm. The goal of controlling and ultimately harvesting these reserves continues to be the driving force behind American interest in this frozen region of the world.

## Arctic Life

The Arctic region is rich in wildlife and this was the first product and the first export from these frozen lands. As discussed, trade in whale oil and seal skins was key in the first generation of Arctic trade. Food fish was the other early pillar of the Arctic's global economic reach, and the fishing of Arctic waters has remained a major source of revenue and livelihood in the region ever since. A 2023 report from the Arctic Economic Council states that there are 633 fish species in the Arctic Ocean that could be harvested for human use, but that only fifty-eight species are currently fished for commercial export. The most recently estimate on the commercial output of Arctic fisheries came in 2016, when it was estimated that Arctic nations exported 5.6 billion kilograms of food into the global seafood market, with an estimated value of $24.8 billion internationally.[2]

Protecting the fishing industry in the Arctic has many dimensions. For one thing, fishing and the export of fish constitutes an essential part of local Arctic economies. Without this industry, many Arctic communities would likely experience significant if not utterly devastating economic turmoil. For Indigenous Arctic residents, the future maintenance of fish stocks is not only economically important, but is essential to the maintenance of cultural traditions as well. Then there are the many species living in the Arctic, including bears, walruses, whales, and sharks that depend on the Arctic food web, of which Arctic fish species harvested for food play an important role.

In considering seafood resources in the Arctic, it is important to understand that climate change is a threat to the stability of fish stocks, but also provides opportunities for those who seek to exploit those stocks. Humanity's access to seafood is declining, in part because of the enormous damage done to ecosystems in which food fish and other seafood organisms are harvested, and, in part, because climate change is reducing the capacity of these environments to sustain life. The retreat of sea ice will provide additional opportunities for companies and individual entrepreneurs seeking to profit from the harvest of Arctic wildlife, but this advantage will be transitory, because the changes to the climate that are driving this environmental transformation will eventually be devastating to the environments that these species need to survive and because growing demand will soon deplete any new sources of oceanic resources that might be discovered thanks to the ravages of climate change.

## Arctic Stones

The search for gold in the Arctic eventually resulted in success, with the discovery of gold in Alaska's Arctic range, resulting in the famed Yukon Gold Rush of the early 1900s. Gold deposits were eventually discovered and mined, across much of Alaska and the state continues to be a major source of gold. The Russian territories in the Arctic also proved to hold gold deposits, and mining continues today in the Chukotka and Yakutia mountains. Other Arctic nations likewise found gold, including in Norway where the Bidjovagge gold mine is still active.

Gold, as valuable and valued as it is, is but a small part of the mineral riches that drew historical explorers and modern mining companies to the Arctic terrain. In addition to gold, copper, nickel, iron ore, and essential elements like bauxite and phosphate are mined in the Arctic region. Russia, thanks to their control over vast portions of the Arctic, has benefitted more notably from the region's mineral wealth. Russian mining companies are leading international producers of nickel, for instance, in part because of their Arctic properties and territories.

Just as with the Arctic's biological resources, climate change is making it easier to access the Arctic's mineral resources as well, and companies have accelerated mining prospecting as new regions become available or become more hospitable to operations. This renewed mineral resource explosion not only refers to the Arctic but also to other areas like offshore mining and the ocean seabed. Naomi Klein, a Canadian environmentalist who wrote the book *This Changes Everything: Capitalism vs. The Climate*, about climate change, uses the term "extractivism" to refer to the push, on behalf of mineral companies, to extract minerals from "sacrifice zones," defining these areas as "places that, to their extractors, somehow don't count and can therefore be poisoned, drained or otherwise destroyed."[3] Mining of minerals (not to mention fossil fuels) causes tremendous amounts of pollution affecting the water, and air, and contaminating landscapes with heavy metal. Sediments and other waste can quickly devastate aquatic and semiaquatic ecosystems. Because these activities are now occurring in areas that are sparsely populated and thus less often the focus of environmentalists, the shift towards this kind of mining and resource extraction meets less resistance, but is no less damaging to the environment in the longer term.[4]

## Arctic Oil

By far the biggest allure of the Arctic to industrialists is Arctic oil deposits. It was in August of 1920 when the first Arctic oil deposits were discovered in the Northwest Territories of Canada, near the McKenzie River. This led to decades of oil speculation, but it wasn't until the 1960s that companies discovered fruitful oil deposits in the Arctic realm. The first was discovered in 1962 by Russian prospectors near Tazovskoye, which became the West Siberian Oilfield. In the United States, billionaire's became excited with oil was discovered in Alaska's Prudhoe Bay in 1968. By 1977, the Trans-Alaska Pipeline had been completed to carry oil from Prudhoe Bay to Valdez, where it could be shipped to the US mainland or to customers overseas. Concern about the oil business, however, peaked after the now infamous *Exxon Valdez* disaster, where an oil tanker crashed, releasing 40 million liters of oil in Alaska's Prince William Sound. It was an ecological nightmare and marked the first time that many Americans had cause to question the ecological impact of the oil industry. How could this be prevented? This is where the debate over Alaska's oil really began to ramp up, and the debate has never ended because there is still oil to collect but collecting it is still an ecological and environmental threat.

Most oil spills attract little, if any, attention. However, the *Exxon Valdez* and the Deepwater Horizon oil spill—resulting from the 2010 sinking of the Deepwater Horizon drilling rig—gained international media attention. Each year, there are more than 150 oil and hazardous chemical spills in US waters. While these spills vary in intensity and impact, all represent a serious ecological hazard with the potential to kill wildlife and to spread illness to human populations as well, including those who must clean up after major oil accidents. A 2022 study, for instance, indicated that workers who cleaned up the British Petroleum (BP) oil spill in 2010 were more than 60 percent more likely than those not involved in the clean-up effort to suffer from asthma and other respiratory disorders.[5] Spilled oil and burning oil means air pollution, and the impact of this pollution spreads far and wide from the area where the accident occurs.

The 19.6-million-acre Arctic National Wildlife Refuge of Alaska, established in 1980, was the result of decades of conservationist efforts. Alaska's wilderness is home to a unique assemblage of endemic species that cannot exist elsewhere, and to wonderful natural vistas representing the Arctic realm. The refuge is home to wolves, polar bears, caribou, and musk oxen, and an annual resource visited by millions of birds representing hundreds of species. To many Americans, this is a resource that should be protected so that at least some remnant of what the Alaskan Arctic once was will be available to future generations of Alaskans and Americans as a whole. The Coastal Plain of the Arctic National Wildlife Refuge is part of the traditional home range of the Gwich'in, an area that these Indigenous Americans call "the sacred place where life begins." Yet there are another group of Americans who feel it would be more profitable and valuable to harvest the fossil fuels that underlie this landscape than to protect the landscape as an entire and unmolested ecosystem. Individuals adopting this view have been trying to bypass or defeat conservation law in an effort to tap into these reserves.

The US government's approach to the Arctic reserve depends on which political party dominates. Though American conservatives were once, as their name suggests, deeply interested in the "conservation" of habitats and America's natural resources, the evolution of conservatism has taken conservatives away from this tradition and has aligned the conservative movement more and more with the interests of corporations and business interests—a focus on direct monetary profit rather than the preservation of resources that might have longer-term but more intangible benefits to humanity and the American people.

Donald Trump and his administration's corporate and wealth-aligned officials epitomize this manifestation of American conservatism, eschewing or ignoring environmental, ecological, and conservation interests in favor of accessing sources of corporate profit. The Trump administration, upon taking office, cancelled federal regulations on air and water safety, for instance, in favor of allowing energy companies to operate without the constraints of having to manage pollution. In 2025, Trump issued an executive order, bypassing even the authority of the legislature, to protect oil companies from having to pay fines when their emission levels exceed state law limits. This focus on economic exploitation has therefore bypassed even conservative belief in state autonomy, expanding central authority and bypassing the will of citizens in states where these laws were established.[6]

The Trump administration's support of the oil industry is justified by the claim that the United States will benefit more from extracting and/or protecting access to petroleum resources than by establishing policies based on more holistic principles of concern. Furthermore, Donald Trump and at least some allies have expressed skepticism about the global scientific consensus on climate change, despite there being no legitimate evidence to suggest that the current understanding is flawed. This climate skepticism then filters into the Trump administration's dismissal of climate-oriented objections to further oil extraction and justifies their efforts to prioritize the shorter-term profit of energy companies over the longer-term preservation of resources and/or investment in alternative and green energy. With regard to the Arctic, the Trump administration has suggested accelerating efforts to drill in the Arctic National Wildlife Refuge, despite the testimony of ecologists, environmentalists, climate scientists, and other experts warning of the potential and expected consequences of doing so. The National Resources Defense Council (NRDC) summarizes the issue, by writing, "The debate over what to do with this landscape has raged for nearly a century, but now, in the midst of a climate crisis that's wreaking havoc at every latitude but warming the poles at astonishing rates, there's broad consensus that drilling the Arctic for fossil fuels is beyond a terrible idea."[7]

## Riches and Risks

As of 2025, the US government has firmly aligned with fossil fuel companies, mineral extraction companies, and the other corporations prioritizing Arctic wealth over Arctic preservation. This brings the United States in line with Russia and other nations that have demonstrated a comparative lack of concern with regard to long-term resource preservation or conservation, in favor of more proximate economic benefit. There are, however, other regional governmental agencies, courts, and nongovernmental organizations (NGOs) that oppose the further exploitation of the Arctic's resources or, at least, that would seek to manage this development in ways that pay attention to the welfare of Indigenous cultures, wildlife, and public health and welfare. If the United States federal government is no longer posed to play a role as an agent of ecological management, than it remains to be seen what influence other governments, NGOs, and the findings of researchers both in the United States and abroad, will have on Arctic development moving forward.

## Works Used

"King Henry III's Polar Bear." *Historic UK*, 2025, www.historic-uk.com/CultureUK/Henry-III-Polar-Bear/.

Klein, Naomi. *This Changes Everything: Capitalism vs. The Climate*. Simon & Schuster, 2014.

"The Long, Long Battle for the Arctic National Wildlife Refuge." *National Resources Defense Council (NRDC)*, 15 Mar. 2024, www.nrdc.org/stories/long-long-battle-arctic-national-wildlife-refuge.

"Oil Spill Cleanup Workers More Likely to Have Asthma Symptoms." *National Institutes of Health (NIH)*, 17 Aug. 2022, www.nih.gov/news-events/news-releases/oil-spill-cleanup-workers-more-likely-have-asthma-symptoms.

Rowe, Mark. "Arctic Nations Are Squaring Up to Exploit the Region's Rich Natural Resources." *Geographical*, 12 Aug. 2022, geographical.co.uk/geopolitics/the-world-is-gearing-up-to-mine-the-arctic.

"The State of Arctic Food." *Arctic Economic Council (AEC)*, 2023, arcticeconomiccouncil.com/wp-content/uploads/2023/09/aec-arctic-food-report-2023.pdf.

"Trump Signs Order to Protect Big Oil from State Emissions Fines." *Bloomberg*, 9 Apr. 2025, www.bloomberg.com/news/articles/2025-04-09/trump-signs-order-to-protect-big-oil-from-state-emissions-fines.

## Notes

1. "King Henry III's Polar Bear," *Historic UK*.
2. "The State of Arctic Food," *Arctic Economic Council (AEC)*, 2023.
3. Klein, *This Changes Everything*.
4. Rowe, "Arctic Nations Are Squaring Up to Exploit the Region's Rich Natural Resources."

5. "Oil Spill Cleanup Workers More Likely to Have Asthma Symptoms," *National Institutes of Health (NIH)*.
6. "Trump Signs Order to Protect Big Oil from State Emissions Fines," *Bloomberg*.
7. "The Long, Long Battle for the Arctic National Wildlife Refuge," *National Resources Defense Council (NRDC)*.

# There's a Global Tug-of-War for Greenland's Resources—But the New Government Has Its Own Plans

By Nicolas Jouan
*The Conversation*, March 24, 2025

Greenland's parliamentary election was held on March 11 against a backdrop of repeated calls from the Trump administration for America to annex the island. The poll delivered a momentous shift in Greenland's political landscape as the pro-business Demokraatit (Democrats) emerged as the biggest winners overturning the two left-leaning parties which had formed the previous government.

Securing nearly 30% of the vote and gaining seven seats for a total of ten in the Inatsisartut (parliament), the party now holds the strongest mandate it has ever had. Close behind was the nationalist Naleraq party, which secured 24.5% of the vote and gained four seats, bringing their total to eight.

While both parties are united in their rejection of Trump's ambitions and share a vision of Greenlandic independence, their approaches couldn't be more different. Demokraatit advocates for a gradual, measured process, prioritising economic development. The party considers that economic self-sufficiency and strengthening domestic infrastructure are key preconditions to achieve independence. Naleraq, on the other hand, is pushing for a rapid break from Denmark. Its line is that Greenland will only be able to unleash its potential, economic and otherwise, once independent.

Independence has long been the dominant theme of Greenlandic politics. Ever since the territory gained home rule in 1979, most political parties across the spectrum have championed the idea of full independence from the kingdom of Denmark. Even the two major challengers—the Inuit Ataqatigiit, which lost five seats at the election to drop to seven, and the once-dominant Siumut, which lost six and now holds just four seats—are pro-independence.

But while independence remains a defining issue, the real story of this election is Greenland's economy. The island is sitting on a treasure trove of rare earth elements, uranium, iron and other minerals critical to global industries. Yet despite decades of interest from foreign investors, strict regulations and environmental concerns have often slowed development.

With Demokraatit's rise, that could change. The party is pushing for pro-business policies, including tax incentives, streamlined regulations and reduced state

From *The Conversation*, March 24 © 2025. Reprinted with permission. All rights reserved.

intervention in key industries like mining, fisheries and tourism. If successful, these reforms could transform Greenland into a major player in the global supply chain.

Despite its electoral gains, Demokraatit faces a challenge in implementing its economic vision. The party's potential coalition partner, Naleraq, is deeply sceptical of foreign investment, at least when it comes from Denmark and Europe. While open to partnerships with the US, Naleraq is adamant that Greenland must retain full control over its resources, resisting any foreign influence that could compromise national sovereignty.

This ideological divide could create friction within a potential coalition government. Will Demokraatit's pro-business agenda be tempered by Naleraq's nationalistic stance? Or will the promise of economic growth push both parties toward compromise?

## Global Powers Are Watching

Greenland's election came at a time when it was already the focus of world attention. Its strategic location and vast resources have attracted growing interest from global superpowers – none more so than the US. Trump has repeatedly expressed interest in acquiring Greenland, a move widely considered unrealistic, but indicative of Washington's strategic priorities.

American interest in Greenland isn't new. The island is home to the Pituffik Space Base, formerly Thule Air Base, since the 1950s as a critical part of North American missile defence and whose Arctic position makes it a key player in both American territorial defence and Nato's security architecture. Pituffik is the only non-Danish military presence in the territory and is the northernmost American military base.

But the White House's rhetoric has taken a more insistent tone, raising questions about whether the US might attempt to exert greater influence over Greenland's economic and political future. The interest in Greenland seems guided by at least two factors: its strategic position at the centre of the North Atlantic security complex and its economic potential with hard-to-access but abundant resources.

In both cases, the growing involvement of both Russia and China in the Arctic seem to make the US wary of a potentially independent Greenland getting closer to unfriendly great powers.

> **The interest in Greenland seems guided by at least two factors: its strategic position at the centre of the North Atlantic security complex and its economic potential with hard-to-access but abundant resources.**

Denmark's central government is walking a diplomatic tightrope when it comes to responding to the US government's repeated intentions to annex Greenland. Copenhagen has sought to Europeanise the debate, floating the idea of Greenland joining the European Union. Taking this step would provide welcome economic support to the island but could also clash with Greenland's scepticism toward European interference.

Greenland now stands at a crossroads. Domestically, negotiations between Demokraatit and Naleraq will likely shape the trajectory of the island's economic and independence ambitions. Internationally, major powers—including the US, the EU and possibly even China and Russia—are positioning themselves to engage with Greenland's untapped potential.

As the world's focus on Greenland intensifies, one thing is clear: this Arctic nation is no longer a remote outpost. It is fast becoming a key battleground for economic, political and strategic influence in the North Atlantic.

**Print Citations**

**CMS**: Jouan, Nicolas. "There's a Global Tug-of-War for Greenland's Resources—But the New Government Has Its Own Plans." In *The Reference Shelf: U.S. National Debate Topic 2025–2026: Exploration & Development in the Arctic,* edited by Micah L. Issitt, 86–88. Amenia, NY: Grey House Publishing, 2025.

**MLA**: Jouan, Nicolas. "There's a Global Tug-of-War for Greenland's Resources—But the New Government Has Its Own Plans." *The Reference Shelf: U.S. National Debate Topic 2025–2026: Exploration & Development in the Arctic,* edited by Micah L. Issitt, Grey House Publishing, 2025, pp. 86–88.

**APA**: Jouan, N. (2025). There's a global tug-of-war for Greenland's resources—But the new government has its own plans. In M. L. Issitt (Ed.), *The reference shelf: U.S. national debate topic 2025–2026: Exploration & development in the Arctic* (pp. 86–88). Grey House Publishing. (Original work published 2025)

# How Trump's Looming Tariffs May Impact Canada's Arctic

By Hilde-Gynn Bye
*High North News*, February 5, 2025

On Monday, US President Donald Trump agreed to pause tariffs on goods from Canada and Mexico for 30 days as the two countries made promises to bolster border security, taking steps to address Trump's concerns about fentanyl trafficking.

The proposed tariffs consisted of 25 percent duties on imports from the neighboring countries, except a lower ten percent tariff on energy products from Canada.

In Canada's vast Arctic regions, officials have made clear the impact that a trade war between the US and Canada would have on the territories in the Canadian North.

## Does Not Resolve the Uncertainty

Ranj Pillai, Premier of Yukon, which borders Alaska in the west, says the recent US tariff delay is a step in the right direction.

The premier noted however that Yukon businesses and communities cannot plan for the future under the constant uncertainty of potential trade barriers.

"While today's news was positive, it does not resolve the uncertainty facing Yukon workers, businesses and communities. Here in the Yukon, our economy depends on fair and predictable trade with American partners," Pillai said in a statement.

"These tariffs—if and when they come into effect—threaten jobs, increase costs for families and disrupt supply chains that have benefited both sides of the border, for decades."

## Impacting the Cost of Living

In a post on social media on Saturday, Premier P.J. Akeeagok of Nunavut, stressed that tariffs may particularly impact the cost of living.

"Nunavummiut [inhabitants of Nunavut, ed. note] are faced with the highest costs of living in Canada, as almost all our goods are flown in from southern jurisdictions," Akeeagok said.

From *High North News*, February 5 © 2025. Reprinted with permission. All rights reserved.

"These tariffs will inevitably put more pressure on families already struggling to pay the rent and feed their children. The tariffs may also impact the cost of building homes and our ability to address our housing crisis."

## Less Exposed, but Will Feel Ripple Effects

In an interview with CBC, the Yukon premier noted that the three territories do not have the same level of exposure as provinces do, but listed some of the Yukon industries that could be hit by tariffs.

"We have folks here that build products for housing and building like Northerm [Windows and Doors] that have subsidiaries in Alaska. We have Hecla [Mining Company], which is exporting materials like silver and lead, and you have to take into consideration what their input is," he told CBC.

Premier of Northwest Territories (NWT), R.J. Simpson, also noted that the territory has limited direct exports to the United States, though adding that the NWT would not be immune to the repercussions of the tariffs.

"Our economy, businesses, and communities will feel the ripple effects," said Simpson.

## Impacts on the Mining Industry

Mining is a large economic driver for the territories in the North, with the mines operating in the Northwest Territories and Nunavut being the largest private sector contributors to each territory's economy, according to a report from the Mining Association of Canada.

Mining company Agnico Eagle's two active mines furthermore account for 22 percent of Nunavut GDP. The company employs over 3500 people, and is the Territory's largest private sector partner.

Agnico Eagle has not responded to *High North News'* request for comments on how potential tariffs would impact the company's operations in Nunavut.

Peter Akman, Head of Communications at Baffinland, which operates the Mary River Mine in Nunavut, said the mining company's high-grade iron ore is currently only shipped to Europe. As such, the company does not foresee an immediate direct impact on their ore sales from tariffs on Canadian exports to the U.S.

"However, any retaliatory tariffs imposed by Canada on imported goods, including machinery, parts or food, or the indirect effect of U.S. tariffs on Canadian components used in the manufacturing of machinery and equipment, would affect our costs," Akman tells *HNN*.

> **Large Economic Driver**
>
> Mining, quarrying and oil and gas extraction were responsible for 15 percent of the GDP in the Yukon, 22 percent of GDP in the Northwest Territories, and 41 percent of GDP in Nunavut.
>
> *From the annual report from the Mining Association of Canada.*

## Concerns about Countervailing Tariffs

"A significant portion of our plants, equipment, and parts are sourced from U.S. based manufacturers. A 25 percent tariff on major components would impact the profitability of our current operations and capital projects. We will be evaluating moving our procurement of these key supplies to Europe and elsewhere," he added.

Akman also noted concerns about the impact of countervailing tariffs announced by the Government of Canada on food imports.

**These tariffs threaten jobs, increase costs for families and disrupt supply chains that have benefited both sides of the border, for decades.**

"Given that much of our food supply for our site workforce comes from or through the U.S., this could increase food prices for us and the residents of Nunavut as a whole. We are working with our suppliers to find alternate sources within Canada and elsewhere."

"We continue to monitor the situation closely to assess any potential financial impact on our supply chain and operations."

## A Time to Invest in the Arctic

In closing, Nunavut Premier Akeeagok noted the need to "do more than react," calling for increased investment in the northern economies.

"Canada's Arctic is a region of opportunity, from critical minerals in our land to an abundance of fish in our waters, it's time to invest in our own economy and communities. In a time of uncertainty, the Arctic can unlock new economic opportunities for our country," he said.

NWT Premier Simpson also highlighted that recent developments present an opportunity to strengthen the diversity of the economy, pointing among other things to promoting tourism, infrastructure projects, extraction of critical minerals and resources, as ways to explore sustainable economic growth.

### Print Citations

**CMS**: Bye, Hilde-Gynn. "How Trump's Looming Tariffs May Impact Canada's Arctic." In *The Reference Shelf: U.S. National Debate Topic 2025–2026: Exploration & Development in the Arctic,* edited by Micah L. Issitt, 89–91. Amenia, NY: Grey House Publishing, 2025.

**MLA**: Bye, Hilde-Gynn. "How Trump's Looming Tariffs May Impact Canada's Arctic." *The Reference Shelf: U.S. National Debate Topic 2025–2026: Exploration & Development in the Arctic,* edited by Micah L. Issitt, Grey House Publishing, 2025, pp. 89–91.

**APA**: Bye, Hilde-Gynn. (2025). How Trump's looming tariffs may impact Canada's Arctic. In M. L. Issitt (Ed.), *The reference shelf: U.S. national debate topic 2025–2026: Exploration & development in the Arctic* (pp. 89–91). Grey House Publishing. (Original work published 2025)

# Critical Minerals in the Arctic: Forging the Path Forward

By Brett Watson, Steven Masterman, and Erin Whitney
*Wilson Center*, July 10, 2023

Critical materials play a vital role in powering modern technologies, from renewable energy systems and electric vehicles to advanced electronics and national defense. However, the United States faces challenges in securing a reliable supply of these materials, which are essential for its economic competitiveness and national security. These materials include metals such as lithium, cobalt, rare earth elements, and graphite. They are essential for the manufacturing of high-tech products like batteries, semiconductors, magnets, and catalysts. With the increasing demand for electric vehicles, renewable energy systems, and advanced electronics, the United States heavily relies on critical materials to drive innovation and maintain its global economic competitiveness.

The United States faces several challenges in securing a stable supply of critical materials. One major challenge is the concentration of production in a few countries, primarily China. China dominates the global supply chain of critical materials, accounting for a significant share of mining, processing, and refining capacity. This dependency raises concerns about supply disruptions, geopolitical risks, and market manipulation.

This problem is complex requiring a multi-pronged approach to develop the best path forward. Attention from multiple agencies with private sector involvement is needed to address issues including resource definition, recovery technology development, recycling technologies, international relations, supply chain resilience and improvements and, domestic resource development policies. A successful outcome will necessitate a diverse approach, strong collaborations, exchange of information and practices, and perhaps most importantly, persistence.

Ensuring a reliable and sustainable supply of these materials is essential for the country's economic growth, technological advancement, and national security. By investing in domestic production, promoting responsible sourcing, and fostering international collaborations, the United States aims to strengthen its critical materials supply chains. These efforts will contribute to a more resilient and secure future, enabling the United States to maintain its position as a global leader in innovation and technology.

This overview explores the issue of critical materials supply chains in the United States, examining the importance of these materials, the challenges asso-

From *Wilson Center*, July 10 © 2023. Reprinted with permission. All rights reserved.

ciated with their supply, efforts being made to address the issue, and research avenues.

## America's Arctic in the Critical Mineral Arena

Arctic regions share many attributes that inspire and challenge resource development. Developing and sharing Arctic-friendly technologies and solutions across international borders, will strengthen Arctic relations, build Arctic economies, help disadvantaged communities and strengthen the nations critical mineral supply chain. Alaska is proud to participate alongside its Arctic colleagues in developing new technologies to meet this challenge.

Alaska's complex geological history has led to formation of a wide array of mineral deposit types containing commodities many list as critical. Alaska either has, is, or could produce almost all of the commodities on the US Geological Survey's 2022 list of critical minerals. Alaska is the largest producer of zinc in the nation, contains the nation's largest graphite deposit, is the state with the only domestic tin resources and, has been a producer of critical minerals in times of national need. e.g. During WWII Alaska contributed tin, PGE's, chrome, tungsten and antimony for the war effort. Most of the commodities produced to support the war effort have not been significantly produced since, and the resources remain in place, creating a ripe environment for meeting the nations need for these critical minerals.

Alaska is large in the domestic context, being about one fifth of the size of the contiguous states. Millions of acres of land selected by native corporations and the state were selected based on mineral potential. These, along with open federal lands provide almost half of Alaska's 400 million acres open to mineral development. Alaska has an active resource development industry known for rigorous environmental and social justice practices.

The University of Alaska is uniquely positioned to assist the national need for a stable critical minerals supply chain, with many of the necessary components to be a center of critical minerals research already in place, including the Mining Engineering and Mineral Processing, Geology, Economics and Chemistry departments, the Mineral Industry Research Laboratory, the unmanned aircraft program, the Geophysical Institute, the Institute of Northern Engineering, a strong relationship with the mineral industry, a strong state geological survey, and strong State of Alaska support for critical mineral research. Centered in the heart of Alaska's mineral industry, the University of Alaska's planned Alaska Critical Mineral Research Center creates potential broad research areas to focus on to facilitate Alaskan, American, and Arctic critical mineral production.

Develop centers of excellence and innovation to research and develop solutions that support stable domestic supplies of critical minerals

## What Is Important Depends on Your Perspective

The perception of what constitutes a critical material is contingent upon one's perspective. Different industries and regions have distinct supply chains and material requirements. For instance, what is considered critical for a solar installation company in one country may not be for a vehicle manufacturer in another. Divergent perspectives arise from variations in supply chains and material needs. In the case of a thin-film solar panel manufacturer, concern might involve securing a stable supply of minerals like indium or tellurium. Conversely, a vehicle manufacturer might prioritize access to critical materials like lithium and cobalt, which are indispensable for the production of electric vehicle batteries. Engaging with industry to determine research needs is crucial to maximize research investment impacts.

> Establish a pathway for industry to engage in public-private partnerships to focus research.

Further, material criticality is dynamic. A material that is critical today may not be in the near future. Technological advancements play a crucial role in changing mineral criticality. For instance, the discovery of the Bayer process in 1888 revolutionized the extraction of aluminum from bauxite, significantly increasing its availability. Geopolitical forces also influence mineral criticality, as shifts in trade relationships, conflicts, and policy changes can impact the accessibility of certain minerals. As technologies advance and new applications emerge, the demand for certain minerals may increase or decrease, leading to shifts in their criticality over time. Potential for rapidly changing demand dynamics creates risk for mineral developers, disincentive to investment, and negatively impacts supply of these minerals.

> Establish domestic mineral policies that mitigate criticality fluctuation risk for mineral investment.

## Supply Chains

The supply chains for critical materials in the United States face potential bottlenecks that extend beyond domestic mining. Even if the United States enhances its domestic mining operations, challenges are present in other stages of the supply chain. These include processing, refining, and manufacturing, which are equally critical for a secure and reliable supply of critical materials.

The case of copper is illustrative. The top four countries in terms of identified reserves of copper are Chile, Australia, Peru and Russia which collectively account for 49% of global reserves. The top four countries in terms of primary mine production are Chile, Congo (Kinshasa), Peru, and China which collectively account for 52% of primary mine production. However, at the refining stage, the top four countries of China, Chile, Congo (Kinshasa), and Japan collectively account for 63% of refined production, with China alone representing 42% of global refining.

> Engage with smelting and refining facilities in the Americas and in geopolitically aligned countries to assess their capacities to recover CM from Alaskan ores, and develop process modification strategies to recover critical minerals not currently recovered.

In terms of processing and refining, the United States relies heavily on foreign countries for these crucial steps. Limited domestic processing facilities and expertise ensure foreign transformation of raw materials into usable forms. Investing in domestic or allied processing capabilities will reduce dependence on foreign processing and mitigate potential disruptions. Engaging with processing facilities in allied countries to expand processing infrastructure and increase metal recovery in those facilities can strengthen the supply chain and reduce supply-chain vulnerability.

## Increasing Domestic Supply

Geologic processes often concentrate several metals of economic interest together in a single deposit. Many critical minerals are produced as by-products in the extraction, processing, and refining of other, more economically significant metals. For example, germanium and indium are mostly produced as a by-product of zinc; gallium a by-product of aluminum, and tellurium is a by-product of copper production. Because the global demand of some critical minerals is small, mines dedicated exclusively to these metals would not be economically sufficient to warrant exploration. Those commodities will still need to be produced as co-products or by-products from production of more economically significant metals.

Baseline geological, geophysical and geochemical data is being acquired by the University of Alaska, the Alaska Division of Geological & Geophysical Surveys (DGGS) and the United States Geological Survey (USGS) through programs like the Department of Energy's (DOE) CORE-CM, and the USGS's Earth MRI. These programs are producing the modern digital baseline data of our sedimentary basins and crystalline rocks that industry uses to evaluate those regions for critical mineral resources. Much remains to be done, and these programs need to continue until domestic

> Develop drone-carried geophysical survey equipment to reduce the cost, increase the speed and density of data collection.

data coverage is complete. Increasing the speed and reducing the cost of data collection through automation may be achieved through drone based-surveys. The University of Alaska is well positioned to research and develop drone-carried geophysical equipment to conduct regional and prospect-scale surveys.

Similarly, researching and developing drone-based hyperspectral surveying capability to make hyperspectral data acquisition and processing more practically available. Incorporating hyperspectral data in regional or prospect-scale surveys to support critical mineral assessments will facilitate resource discovery and development. One limitation with hyperspectral data is its limited utility in vegetated areas. Developing algorisms to filter vegetation signals from the data will broaden the usability of this method.

Modern geological data is geospatial and digital. Machine learning (ML), or artificial intelligence (AI) algorithms could leverage this digital geophysical, geochemical and geological data to predict locations of interest for critical minerals.

These would allow national or state entities and industry to focus resources on highest potential areas and shorten the time to discovery and production.

Develop AI or ML routines using a mineral systems approach to predict CM locations from integrated digital data.

High transportation and energy costs in remote arctic regions are a significant economic barrier to development. The costs for transporting concentrates for further processing and metal extraction at a distant facility can be insurmountable for remote locations. Process technologies that allow shipment of metals rather than ore from remote locations would greatly improve project economics, and increase domestic production. At the same time, many conventional refining and metal extraction processes are energy intensive, making them prohibitively expensive in remote Arctic areas that face some of the nation's highest energy costs. Processes which improve recovery efficiency or concentrate grade without requiring more energy have the greatest potential to enhanced value added activity. As an example, in the gold mining industry bio-oxidation of sulfides followed by gold leaching is a method to recover gold from appropriate ores. Extending this or similar process technology to appropriate critical mineral ores could significantly improve the economics of remote deposits, bringing more into production.

> Support metallurgical research to increase ability for onsite reduction to metals.

Industry investment can be incentivized by providing baseline metallurgical information that demonstrates recovery viability for deposits in a region. This can be achieved by conducting regional characterization studies for critical mineral deposits to determine whether the mineralogy is amenable to metallurgical recovery, and whether the ores contain deleterious elements that would impact economics. For example, do any of Alaska's belts of antimony deposits contain deleterious elements that would be barriers to development?

Advanced exploration and development projects focus metallurgical work on the dominant commodities in an orebody, frequently paying little to no attention to minor commodities that would be a by-product. Engaging with projects in the advanced exploration to development phase to characterize critical mineral content, and developing solution to production barriers would help facilitate critical mineral production from the next generation of domestic mines. For example, cobalt is contained in the pyrite at the Bornite deposit in northwest Alaska, what metallurgical solutions can be developed to facilitate its recovery.

> Support mineral-system based metallurgical research to characterize regional belts of mineral deposits.

Barriers to critical mineral production must exist at current mining operations where the contained critical minerals are not being produced. Determining these barriers and researching solutions to facilitate critical mineral production where appropriate could bring rapid results, e.g., ore at the Greens Creek and Red Dog mines contains significant barite, but neither mine produces barite as a salable product. By contrast the Palmer project, which is geologically similar to these deposits, is planning to produce barite as a salable product.

Process waters from oil and gas production potentially contain critical minerals. Development of carbon resources that result in metal-rich brine production from organic deep-water sediments offer potential for by-product critical mineral production. On Alaska's North Slope some of the oil and gas source rocks are known to contain high REE, V, P and fluorine. Extraction techniques developed to extract critical minerals from Alaskan production waters could be utilized across the Arctic. Produced water and reservoir composition datasets may allow al mineral prediction of critical mineral content in undeveloped unconventional reservoir.

> Engage with industry to develop recovery circuits for critical minerals at existing and developing mines.

## Reducing Import Levels

The demand for metals is expected to growth significantly between now and 2030. The rapid industrialization in emerging economies, particularly in Asia, will contribute to increased demand for metals. The transition towards renewable energy sources, including wind and solar power, will also spur the demand for metals like lithium, cobalt, and rare earth elements used in the production of batteries, permanent magnets, and energy storage systems. E.g., the International Monetary Fund forecast that graphite production will need to increase by almost 1000% to meet the need for electric vehicle batteries.

Looking ahead to 2050, the growth in demand for metals is projected to continue, driven by transformative global trends. The global population is expected to reach nearly 10 billion by 2050, leading to increased consumption. The continued shift towards sustainable energy systems and the electrification of various sectors, including transportation and industrial processes, will sustain the demand for metals.

Without increased domestically available supplies, increased global demand for these commodities will deepen the nation's exposure to an insecure supply chain, and threaten the domestic conversion to renewable energy. The International Monetary Fund concluded that without increased metal availability, demand-based commodity inflation alone could negatively impact the energy transition.

> Develop new recycling solutions to reduce the risks and impacts from global increases in metal demand.

Rapid development of recycling technologies will be crucial to both reduce the environmental impact from increased use of CMs, and reduce the need for new mineral production to meet this demand. Electric car batteries are a clear example of the need for enhanced recycling technologies, with the number of batteries in production, being used and being disposed of projected to rapidly increase with the transition to electric vehicles.

## Domestic Mineral and Economic Policies

Global markets are small for some critical mineral commodities, in terms of the tons of mineral consumed or the dollar value of the global demand, e.g., tellurium. These small market commodities are easily dominated by a handful of producers, or single nation, making them susceptible to price manipulation and supply constraints. Price manipulation creates a particular risk to mineral investors who need assurance that minerals can be produced profitably prior to investing capital in exploration or development. This investment risk nominally increases as the market size of the commodity and number of producers decreases. Risks associated with small market resource development include commodity substitution, new technologies reducing demand, recycling improvements, production technology improvements and, market manipulation and inelasticity. Risks associated with small market commodities deter investment by the minerals industry. National or international policy agreements that mitigate these risks will increase mineral investment and improve critical mineral availability.

> Conduct economic analyses and develop economic and regulatory policy option for incentivizing critical mineral development in rural areas near disadvantaged communities.

> Develop national or international policies or incentives to de-risk market manipulation will spur industry investment in critical mineral development.

> Develop import regulations requiring critical minerals imports meet US environmental and social justice standards.

The environmental and social impacts associated with mining and processing these materials present sustainability challenges that need to be addressed. Environmental regulations in Arctic nations are often more rigorous that in less developed countries. This creates uneven economic conditions favoring development in regions with less stringent environmental and social policies. This can be rectified by developing import regulations requiring critical minerals imports meet rigorous environmental and social justice standards....

### Print Citations

**CMS**: Watson, Brett, Steven Masterman, and Erin Whitney. "Critical Minerals in the Arctic: Forging the Path Forward." In *The Reference Shelf: U.S. National Debate Topic 2025–2026: Exploration & Development in the Arctic*, edited by Micah L. Issitt, 92–98. Amenia, NY: Grey House Publishing, 2025.

**MLA**: Watson, Brett, Steven Masterman, and Erin Whitney. "Critical Minerals in the Arctic: Forging the Path Forward." *The Reference Shelf: U.S. National Debate Topic 2025–2026: Exploration & Development in the Arctic*, edited by Micah L. Issitt, Grey House Publishing, 2025, pp. 92–98.

**APA**: Watson, B., Masterman, S., & Whitney, E. (2025). Critical minerals in the Arctic: Forging the path forward. In M. L. Issitt (Ed.), *The reference shelf: U.S. national debate topic 2025–2026: Exploration & development in the Arctic* (pp. 92–98). Grey House Publishing. (Original work published 2023)

# Unleashing Alaska's Extraordinary Resource Potential

By Donald Trump
*The White House*, January 20, 2025

By the authority vested in me as President by the Constitution and the laws of the United States of America, it is hereby ordered:

Section 1. Background. The State of Alaska holds an abundant and largely untapped supply of natural resources including, among others, energy, mineral, timber, and seafood. Unlocking this bounty of natural wealth will raise the prosperity of our citizens while helping to enhance our Nation's economic and national security for generations to come. By developing these resources to the fullest extent possible, we can help deliver price relief for Americans, create high-quality jobs for our citizens, ameliorate our trade imbalances, augment the Nation's exercise of global energy dominance, and guard against foreign powers weaponizing energy supplies in theaters of geopolitical conflict.

Unleashing this opportunity, however, requires an immediate end to the assault on Alaska's sovereignty and its ability to responsibly develop these resources for the benefit of the Nation. It is, therefore, imperative to immediately reverse the punitive restrictions implemented by the previous administration that specifically target resource development on both State and Federal lands in Alaska.

Sec. 2. Policy. It is the policy of the United States to:

(a) fully avail itself of Alaska's vast lands and resources for the benefit of the Nation and the American citizens who call Alaska home;

(b) efficiently and effectively maximize the development and production of the natural resources located on both Federal and State lands within Alaska;

(c) expedite the permitting and leasing of energy and natural resource projects in Alaska; and

(d) prioritize the development of Alaska's liquified natural gas (LNG) potential, including the sale and transportation of Alaskan LNG to other regions of the United States and allied nations within the Pacific region.

Sec. 3. Specific Agency Actions. (a) The heads of all executive departments and agencies, including but not limited to the Secretary of the Interior; the Secretary of Commerce, acting through the Under Secretary of Commerce for Oceans and Atmosphere; and the Secretary of the Army acting through the Assistant Secretary of the Army for Public Works, shall exercise all lawful authority and discretion available to them and take all necessary steps to:

---

From the *White House*, January 20 © 2025. Reprinted with permission. All rights reserved.

(i) rescind, revoke, revise, amend, defer, or grant exemptions from any and all regulations, orders, guidance documents, policies, and any other similar agency actions that are inconsistent with the policy set forth in section 2 of this order, including but not limited to agency actions promulgated, issued, or adopted between January 20, 2021, and January 20, 2025; and

(ii) prioritize the development of Alaska's LNG potential, including the permitting of all necessary pipeline and export infrastructure related to the Alaska LNG Project, giving due consideration to the economic and national security benefits associated with such development.

(b) In addition to the actions outlined in subsection (a) of this section, the Secretary of the Interior shall exercise all lawful authority and discretion available to him and take all necessary steps to:

(i) withdraw Secretarial Order 3401 dated June 1, 2021 (Comprehensive Analysis and Temporary Halt on All Activities in the Arctic National Wildlife Refuge Relating to the Coastal Plain Oil and Gas Leasing Program);

(ii) rescind the cancellation of any leases within the Arctic National Wildlife Refuge, other than such lease cancellations as the Secretary of the Interior determines are consistent with the policy interests described in section 2 of this order, initiate additional leasing through the Coastal Plain Oil and Gas Leasing Program, and issue all permits, right-of-way permits, and easements necessary for the exploration, development, and production of oil and gas from leases within the Arctic National Wildlife Refuge;

(iii) rescind the final supplemental environmental impact statement entitled "Coastal Plain Oil and Gas Leasing Program Supplemental Environmental Impact Statement," which is referred to in "Notice of Availability of the Final Coastal Plain Oil and Gas Leasing Program Supplemental Environmental Impact Statement, Alaska" 89 *Fed. Reg.* 88805 (November 8, 2024);

(iv) place a temporary moratorium on all activities and privileges granted to any party pursuant to the record of decision signed on December 8, 2024, entitled "Coastal Plain Oil and Gas Leasing Program Record of Decision," which is referred to in "Notice of Availability of the Record of Decision for the Final Supplemental Environmental Impact Statement for the Coastal Plain Oil and Gas Leasing Program, Alaska," 89 *Fed. Reg.* 101042 (December 13, 2024), in order to review such record of decision in light of alleged legal deficiencies and for consideration of relevant public interests, and, as appropriate, conduct a new, comprehensive analysis of such deficiencies, interests, and environmental impacts;

> **By developing these resources to the fullest extent possible, we can help deliver price relief for Americans, create high-quality jobs for our citizens, ameliorate our trade imbalances, augment the Nation's exercise of global energy dominance, and guard against foreign powers weaponizing energy supplies in theaters of geopolitical conflict.**

(v) reinstate the final environmental impact statement entitled "Final Environmental Impact Statement for the Coastal Plain Oil and Gas Leasing Program," which is referred to in "Notice of Availability," 84 *Fed. Reg.* 50472 (September 25, 2019);

(vi) reinstate the record of decision signed on August 21, 2020, entitled "Coastal Plain Oil and Gas Leasing Program Record of Decision," which is referred to in "Notice of 2021 Coastal Plain Alaska Oil and Gas Lease Sale and Notice of Availability of the Detailed Statement of Sale," 85 *Fed. Reg.* 78865 (December 7, 2020);

(vii) evaluate changes to, including the potential recission of, Public Land Order 5150, signed by the Assistant Secretary of the Interior on December 28, 1971, and any subsequent amendments, modifications, or corrections to it;

(viii) place a temporary moratorium on all activities and privileges granted to any party pursuant to the record of decision signed on June 27, 2024, entitled "Ambler Road Supplemental Environmental Impact Statement Record of Decision," which is referred to in "Notice of Availability of the Ambler Road Final Supplemental Environmental Impact Statement, Alaska," 89 *Fed. Reg.* 32458 (April 26, 2024), in order to review such record of decision in light of alleged legal deficiencies and for consideration of relevant public interests and, as appropriate, conduct a new, comprehensive analysis of such deficiencies, interests, and environmental impacts; and reinstate the record of decision signed on July 23, 2020, by the Bureau of Land Management and United States Army Corps of Engineers entitled "Ambler Road Environmental Impact Statement Joint Record of Decision," which is referred to in "Notice of Availability of the Record of Decision for the Ambler Mining District Industrial Access Road Environmental Impact Statement," 85 *Fed. Reg.* 45440 (July 28, 2020);

(ix) rescind the Bureau of Land Management final rule entitled "Management and Protection of the National Petroleum Reserve in Alaska," 89 *Fed. Reg.* 38712 (May 7, 2024);

(x) rescind any guidance issued by the Bureau of Land Management related to implementation of protection of subsistence resource values in the existing special areas and proposed new and modified special areas in the National Petroleum Reserve in Alaska, as published on their website on January 16, 2025;

(xi) facilitate the expedited development of a road corridor between the community of King Cove and the all-weather airport located in Cold Bay;

(xii) place a temporary moratorium on all activities and privileges granted to any party pursuant to the record of decision signed on April 25, 2022, entitled "National Petroleum Reserve in Alaska Integrated Activity Plan Record of Decision," (NEPA No. DOI-BLM-AK-R000-2019-0001-EIS), in order to review such record of decision in light of alleged legal deficiencies and for consideration of relevant public interests and, as appropriate, conduct a new, comprehensive analysis of such deficiencies, interests, and environmental impacts;

(xiii) rescind the Bureau of Land Management final rule entitled "Management and Protection of the National Petroleum Reserve in Alaska," 89 *Fed. Reg.*

38712 (May 7, 2024), and rescind the Bureau of Land Management notice entitled "Special Areas Within the National Petroleum Reserve in Alaska," 89 *Fed. Reg.* 58181 (July 17, 2024);

(xiv) reinstate Secretarial Order 3352 dated May 17, 2017 (National Petroleum Reserve – Alaska), which is referred to in "Final Report: Review of the Department of the Interior Actions that Potentially Burden Domestic Energy," 82 *Fed. Reg.* 50532 (November 1, 2017), and the record of decision signed on December 31, 2020, entitled "National Petroleum Reserve in Alaska Integrated Activity Plan Record of Decision," which is referred to in "Notice of Availability of the National Petroleum Reserve in Alaska Integrated Activity Plan Final Environmental Impact Statement," 85 *Fed. Reg.* 38388 (June 26, 2020);

(xv) reinstate the following Public Land Orders in their original form:

a. Public Land Order No. 7899, signed by the Secretary of the Interior on January 11, 2021;

16. Public Land Order No. 7900, signed by the Secretary of the Interior on January 16, 2021;

17. Public Land Order No. 7901, signed by the Secretary of the Interior on January 16, 2021;

18. Public Land Order No. 7902, signed by the Secretary of the Interior on January 15, 2021;

19. Public Land Order No. 7903, signed by the Secretary of the Interior on January 16, 2021; and

20. any other such Public Land Order that the Secretary of the Interior determines would further the policy interests described in section 2 of this order.

(xvi) immediately review all Department of the Interior guidance regarding the taking of Alaska Native lands into trust and all Public Land Orders withdrawing lands for selection by Alaska Native Corporations to determine if any such agency action should be revoked to ensure the Department of the Interior's actions are consistent with the Alaska Statehood Act of 1958 (Public Law 85-508), the Alaska National Interest Lands Conservation Act (ANILCA) (16 U.S.C. 3101 *et seq.*), the Alaska Native Claims Settlement Act of 1971 (43 U.S.C. 1601, *et seq.*), the Alaska Land Transfer Acceleration Act (Public Law 108-452), and the Alaska Native Vietnam-era Veterans Land Allotment Program under section 1629g-1 of title 43, United States Code.

(xvii) rescind the record of decision "Central Yukon Record of Decision and Approved Resource Management Plan," signed on November 12, 2024, which is referred to in "Notice of Availability of the Record of Decision and Approved Resource Management Plan for the Central Yukon Resource Management Plan/Environmental Impact Statement, Alaska," 89 *Fed. Reg.* 92716 (November 22, 2024);

(xviii) reimplement the draft resource management plan and environmental impact statement referenced in the National Park Service notice entitled "Notice of Availability for the Central Yukon Draft Resource Management Plan/Environmental Impact Statement, Alaska," 85 *Fed. Reg.* 80143 (December 11, 2020);

(xix) rescind the National Park Service final rule entitled "Alaska; Hunting and Trapping in National Preserves," 89 *Fed. Reg.* 55059 (July 3, 2024), and reinstate the National Park Service final rule entitled "Alaska; Hunting and Trapping in National Preserves," 85 *Fed. Reg.* 35181 (June 9, 2020), in its original form;

(xx) deny the pending request to the United States Fish and Wildlife Service to an establish indigenous sacred site in the Coastal Plain of the Arctic National Wildlife Refuge;

(xxi) immediately conduct a review of waterways in the State of Alaska and direct the Bureau of Land Management, in consultation with the State of Alaska, to provide recommendations of navigable waterways subject to the equal footing doctrine and the Submerged Lands Act of 1953, as amended, 43 U.S.C. 1301 *et seq.*, and prepare Recordable Disclaimers of Interest pursuant to section 315 of the Federal Land Policy and Management Act of 1976, 43 U.S.C. 1745, to restore ownership of said waterways to the State as appropriate;

(xxii) direct all bureaus of the Department of the Interior to consider the Alaskan cultural significance of hunting and fishing and the statutory priority of subsistence management required by the ANILCA, to conduct meaningful consultation with the State fish and wildlife management agencies prior to enacting land management plans or other regulations that affect the ability of Alaskans to hunt and fish on public lands, and to ensure to the greatest extent possible that hunting and fishing opportunities on Federal lands are consistent with similar opportunities on State lands; and

(xxiii) identify and assess, in collaboration with the Secretary of Defense, the authorities and public and private resources necessary to immediately achieve the development and export of energy resources from Alaska—including but not limited to the long-term viability of the Trans-Alaska Pipeline System and the associated Federal right-of-way as an energy corridor of critical national importance—to advance the Nation's domestic and regional energy dominance, and submit that assessment to the President.

(c) In addition to the actions outlined in subsection (a) of this section, the Secretary of Agriculture shall place a temporary moratorium on all activities and privileges authorized by the final rule and record of decision entitled "Special Areas; Roadless Area Conservation; National Forest System Lands in Alaska," 88 *Fed. Reg.* 5252 (January 27, 2023), in order to review such rule and record of decision in light of alleged legal deficiencies and for consideration of relevant public interests and, as appropriate, conduct a new, comprehensive analysis of such deficiencies, interests, and environmental impacts. Further, the Secretary of Agriculture shall reinstate the final rule entitled "Special Areas; Roadless Area Conservation; National Forest System Lands in Alaska," 85 *Fed. Reg.* 68688 (October 29, 2020).

(d) In addition to the actions outlined in subsection (a) of this section, the Secretary of the Army, acting through the Assistant Secretary of the Army for Civil Works, shall render all assistance requested by the Governor of Alaska to facilitate the clearing and maintenance of transportation infrastructure, consistent with applicable law. All such requests for assistance shall be transmitted to the Secretary of Defense, Secretary of the Interior, and Assistant to the President for Economic Policy for approval prior to initiation.

(e) The Assistant Secretary of the Army for Civil Works, under the direction of the Secretary of the Army, shall immediately review, revise, or rescind any agency action that may in any way hinder, slow or otherwise delay any critical project in the State of Alaska.

(f) The Secretary of Commerce, in coordination with the Secretary of the Interior, shall immediately review, revise or rescind any agency action that may in any way hinder, slow or otherwise delay any critical project in the State of Alaska.

Sec. 4. General Provisions. (a) Nothing in this order shall be construed to impair or otherwise affect:

(i) the authority granted by law to an executive department or agency, or the head thereof; or

(ii) the functions of the Director of the Office of Management and Budget relating to budgetary, administrative, or legislative proposals.

(b) This order shall be implemented consistent with applicable law and subject to the availability of appropriations.

(c) This order is not intended to, and does not, create any right or benefit, substantive or procedural, enforceable at law or in equity by any party against the United States, its departments, agencies, or entities, its officers, employees, or agents, or any other person.

## Print Citations

**CMS**: Trump, Donald. "Unleashing Alaska's Extraordinary Resource Potential." In *The Reference Shelf: U.S. National Debate Topic 2025–2026: Exploration & Development in the Arctic,* edited by Micah L. Issitt, 99–104. Amenia, NY: Grey House Publishing, 2025.

**MLA**: Trump, Donald. "Unleashing Alaska's Extraordinary Resource Potential." *The Reference Shelf: U.S. National Debate Topic 2025–2026: Exploration & Development in the Arctic,* edited by Micah L. Issitt, Grey House Publishing, 2025, pp. 99–104.

**APA**: Trump, D. (2025). Unleashing Alaska's extraordinary resource potential. In M. L. Issitt (Ed.), *The reference shelf: U.S. national debate topic 2025–2026: Exploration & development in the Arctic* (pp. 99–104). Grey House Publishing. (Original work published 2025)

# The Arctic Trilemma: The United States Must Compete in the Transpolar Sea Route

By Ashton Basak
*Georgetown University Center for Security Studies*, October 5, 2024

Maritime trade channels essential for America's globalized economy lie in the crossfire of security threats. Examples abound: in the Middle East, Iran has seized ships passing through the Strait of Hormuz and has employed Houthi rebels as proxies in the Bab el-Mandeb Strait to target commercial vessels and disrupt trade. The Suez Canal is threatened by the spillover of regional instability with the Israel-Hamas war, while the Strait of Malacca is dominated and increasingly militarized by Beijing. The common denominator is clear: global trade routes are under siege and the United States must look for feasible alternatives.

The United States may turn its attention to the North for alternative trade routes: The Northern Sea Route (NSR) and Northwest Passage (NWP). These trade routes are gaining strategic relevance with the ever faster recession of the Arctic ice shield. The United States is an Arctic power; its direct access to the region and national security objectives in the Arctic give the United States a say in the region's future. However, when it comes to the NSR and NWP, passages used and administered by Russia and Canada, the U.S. is dependent on international cooperation and can only access the existing Arctic trade routes through international partnerships. While these routes are already faster than the Suez or Panama Canal, ice melt from global warming will open an even more efficient Transpolar Sea Route (TSR) by mid-century. This presents a unique geostrategic opportunity, for which many nations will compete. The United States must enter this competition and seek to capitalize on the economic potential of trans-polar trade. Russia already has a formidable military presence in the region and has a major economic and strategic interest in the TSR. Moscow continues to push military capabilities closer to the North Pole, laying claim to the TSR before it opens up ice-free for international use in the next few decades.

The threat of Russia cutting off the United States from the TSR is an economic, strategic, and security challenge that has received too little attention in Washington, DC. If the United States does not stifle encroaching Russian influence in the region now, it might lose its freedom of commercial navigation in the TSR when it opens up in the 2040s—the current best estimate, based on the trajectory of ice melting.

From *Georgetown University Center for Security Studies*, October 5 © 2024. Reprinted with permission. All rights reserved.

The United States can shape its engagement in the region in several ways, however each strategy will have trade-offs. Broadly speaking, the U.S. may have three objectives in the region: (1) the avoidance of conflict with, and, where possible, cooperation with Russia, (2) economic dominance of the TSR, and (3) greater military presence near the TSR. Between three possible outcomes and combinations thereof, only two are ever attainable, with the third always becoming a casualty. The U.S. thus faces an Arctic Trilemma.

What is the way forward? I argue that the United States must embrace Great Power Competition and contain Russia's ambitions in the Arctic by pursuing (2) economic dominance of and (3) military presence in the TSR. The United States must secure dominance of the TSR through military command of the sea, and restructure military plans on the strategic, operational, and tactical levels to contest Russian hegemony in the Arctic.

## The Current Geoeconomic Situation in the Arctic

Trade in the region is economically existential for Russia. Currently, 30% of Russia's GDP can be traced back to the Arctic, and thus, Russia has a vested interest in securing and developing its commercial assets in the region. However, it has not always been this way. For centuries, Indigenous settlers, European explorers, and nation-states have used two conventional Arctic routes for military, economic, and tourist activities: the NSR, which hugs the northern coast of Russia, and the NWP, which rides the North American border and weaves into the Canadian Arctic Archipelago. The NSR was first used by the Soviet Union to extract resources and move military equipment from one side of the nation to the other. The fall of the Soviet Union initially diminished Russian presence in the Arctic, however Putin's rise precipitated renewed interest in the NSR. Trade volumes increased by 755% between 2014 and 2022, and Russia plans to further increase trade volumes tenfold by 2035. As Russia's trade security diminishes in the Black Sea due to its war in Ukraine, the Arctic is becoming an increasingly attractive substitute. Furthermore, a ship traveling between Europe and East Asia takes roughly 30 days to complete the journey via the Suez Canal but this journey can be reduced to 18 days via the NSR.

Canada is another, yet (much) friendlier geopolitical actor in the region with control over the NWP, which runs opposite to Russia's NSR. Dangerous ice coverage on the NWP made passage impossible until the latter 20th century. Due to global ice melting, induced by accelerating global warming, the summer of 2007 marked the first time the NWP could be fully utilized without the concern of icebergs imperiling ships. While Canada grants access to the United States and its partners, it claims that the NWP lies in its territorial waters, as Russia does with the NSR. However, the United States has contested this claim since the 1980s: it posits that under the United Nations Convention on the Law of the Sea (UNCLOS), the NWP is an international strait as the nexus between the Atlantic and Pacific Oceans. While the United States presents the same argument against Russia's NSR, cordial relations with Canada mean that the U.S. need not contest

Canada's claims nearly as fiercely as the Russian ones. However, if the U.S. presses Canada too hard and continues to levy demands against its Northern neighbor, the relationship between the two states may take a toll. Souring cooperation with a continental ally would cost the United States the benefits of shorter commercial trips—shaving up to 2,500 nautical miles off total travel time. The opening of the TSR will simply offer a more economically viable trade route.

## The Opening of the Transpolar Sea Route

The Arctic is warmer than has been at any point over the past 2,000 years. This region is particularly vulnerable to climate change, heating nearly four times faster than the rest of the Earth since 1979. As a result, from 1979 to 2010, Arctic ice coverage has fallen by 40%, while the thickness of the ice has fallen by 70%. Disproportionate and intense warming is changing the geographical structure of the Arctic.

By 2040, this change will open a third maritime trade route called the Transpolar Sea Route (TSR). The route runs from the North Pacific Ocean to the North Atlantic Ocean, cutting through the North Pole. The TSR is different from the NSR and NWP because the bulk of the TSR is not covered by exclusive economic zones or sovereign territorial waters. Furthermore, the TSR is undeniably faster. A ship that uses the TSR can travel through the Arctic one to five days faster than if it used the NSR, while a ship that normally relies on the Suez Canal could save up to 41% of its distance. Shorter travel times means a ship can make more commercial trips, use less fuel, and incur lower operating costs, thereby generating increased revenue. The TSR is the optimal trade route in the Arctic.

While the TSR is not open yet, as the Arctic continues to warm, the Central Arctic Ocean (CAO) could be navigable by ice-breaking cargo ships as early as 2030. By 2045, moderate global warming models predict low enough ice levels in the CAO for standard cargo ships to conduct commercial activity during summer. The question is not if the CAO, and by extension the TSR, will open, but rather, how soon? Given the economic benefits of the TSR, the United States must take steps *now* to secure access for its ships before its adversaries.

## The Threat Landscape

While the TSR may not fully open for another 20 years, Russia has already taken steps to cement its influence in the region. In July 2022, Russian President Vladimir Putin declared the Russian military will defend its claims to the Arctic "by all means" necessary. Putin has backed his rhetoric with substantive commitment to the area: Russia has about 33% more military bases in the Arctic Circle than NATO does. While this may be unsurprising because 53% of the Arctic coastline is under Russian control, that does not mean that the United States and NATO partners should accept a position of regional inferiority.

Russia has more icebreakers—valuable military assets that help clear icy paths for commercial and military ships—than NATO. The U.S. only accounts for two icebreakers in NATO's undersized fleet, lagging far behind its adversary. Icebreakers will be essential in making use of the TSR as early as the 2030s. This may seem relatively far away. But the race begins now. Icebreakers need to be constructed. Military bases need to be built. Transit infrastructure needs to be set up. All of this takes time, and for the U.S. to be ready when the Arctic becomes accessible, it must lay the groundwork now.

Given Russia's military bases and icebreakers, scholars believe that the West trails 10 years behind Russia militarily in the region. Russia has established a foothold near the CAO that will give it oversight, control and area-denial capabilities over the TSR. The U.S. must contest the aspiring Russian hegemony in order to better position the United States toward active use of the TSR.

The potential for Sino-Russian collaboration in the region is concerning. In 2018, China claimed to be a "near-Arctic state," signaling that it seeks to shape the political and economic future of the region. Beijing's goal for a Polar Silk Road that connects Russia to the Belt and Road trade initiative has led to billions of dollars in foreign direct investment. Even more concerning is the aggressive rhetoric with which the Chinese military frames its regional aspirations. Beijing emphasizes that the Arctic represents a "scramble for new strategic spaces," and thus, China "cannot rule out the possibility of using force." This is particularly relevant to the TSR: in the summer of 2020, the Chinese icebreaker MV Xue Long 2 voyaged the TSR for the first time. It would be a mistake to think that Russia is the only adversary to U.S. economic interests in the TSR.

## The Arctic Trilemma: The United States' Way Forward

Given the regional and strategic complexities outlined above, what should the U.S. policy be, and why? To recall, the Arctic trilemma has three legs: access to the TSR, a greater U.S. military presence in the region, and stability and cooperation vis-a-vis Russia. One may consider each side of the trilemma. First, if the U.S. wants free economic access to the TSR but wants to avoid a hostile economic environment by cooperating with Russian interests, it must sacrifice establishing a greater military presence in the region. Russia will not unconditionally cooperate and promote an uncontested peace if the U.S. military challenges their positions in the region. Putin's invasion of Ukraine provides telling evidence: the authoritarian has asserted that he can and will respond forcefully to perceived U.S. and NATO threats to Russian sovereignty. As a result, the U.S. would have to concede to Russian oversight to ensure cooperation and economic access, a position which Washington should—and would—never accept.

Alternatively, the U.S. can have primary access to the TSR and advance military objectives in the region, but sacrifice cooperation with Russia and prompt Great Power Competition in the Arctic. The Russians already have an entrenched military presence in the Arctic and can leverage its assets to dominate the TSR if they are uncontested. Thus, the United States must cultivate a formidable military force in the region to ensure that U.S. companies can utilize the TSR to their maximum economic advantage. Open sea lanes would have to be secured through U.S. and Allied sea control. The competition to be the economic hegemon of the TSR makes any opportunity for joint efforts toward mutual prosperity unlikely and military steadfastness a necessity.

Lastly, the U.S. can maintain peace and project a strong military footprint but suffer from suboptimal access to the TSR. The political climate of the Arctic would resemble a Cold War-esque gridlock due to military build-up by both nations and potentially others who wish to enter the mix. The militarization of the Arctic would sacrifice the greater economic gains both sides could have achieved through a demilitarized strait.

Within this trilemma, the U.S. must secure access to the TSR and establish a strong military presence in the Arctic, embracing great power competition. Given Putin's militarism and completely shattered relations with the West post-February 2022, Washington should not pursue cooperation. To secure U.S. economic interests in the TSR in the 2040s is to ensure no adversary will threaten U.S. ships. Naval theorist Alfred Thayer Mahan argued that securing access to the high seas for trade and commerce is crucial for a nation. Sea lines like the TSR are the economic arteries of Great Powers. They sustain economic life. And they are too important to be left under adversarial control. The U.S. must command the Arctic sea, like it has commanded the Pacific and other oceans for the better part of half a century.

## Strategic, Operational, and Tactical Restructuring

To achieve its foreign policy objectives in and around the TSR, the United States must restructure its approach on the strategic, operational, and tactical level. To frame this section, I will define the strategic level of military planning in the region as how the United States employs its differentiated instruments of power to achieve its objectives in the Arctic. Furthermore, the tactical level deals with the precise arrangement of military forces in the Arctic, while the operational level of military planning is the impetus, holding these asset placements to the standard of the Arctic strategy.

At the strategic level, the United States must construct new air and naval bases, modernize existing ones, or leverage relationships with partners for joint-access in order to close the base race with Russia. Access to the region will be important for air and naval policing missions from a vantage point near the North Pole. In June 2022, the United States was granted unrestrained access to Evenes Air Station and Ramsund Naval Base in Norway, while in February 2024, the United States gained access to eight Norwegian military facilities including Andøya and Bardufoss Air Stations. The United States should continue working with its Norwegian and British allies to develop Evenes, Andøya, and Bardufoss as P-8 surveillance aircraft bases and F-35 hubs for surveillance runs in the Arctic. P-8s are maritime patrol aircraft that can engage in anti-submarine/anti-surface warfare and simultaneously collect intelligence, while F-35s are stealth fighter aircraft that can fly up to supersonic speeds. These two military assets are technological powerhouses that will bolster U.S. military presence in the region if they can be stationed there permanently. The U.S. Navy's Sixth Fleet should also move to be co-located at Ramsund Naval Base, which is used as a major logistics hub for NATO ships and can serve as headquarters for U.S. naval operations in the Arctic. To oversee the other side of the Greenland-Svalbard-Norway (GSN) gap in the TSR, the United States should seek a defense agreement granting the U.S. Air Force permission to utilize the air base in Alert, Canada.

At the operational level, the United States should consolidate military leadership of the Arctic under one Combatant Command: U.S. European Command (USEUCOM). While base expansion and development provides an overall framework for the military in the region, directing the instruments of power at play relies on operational leadership. USEUCOM should assume the military management role of the Arctic by being in charge of coordinating tactical forces to meet the strategic goals. The current problem is that the U.S. Northern Command (USNORTHCOM), U.S. European Command, and U.S. Indo-Pacific Command (USINDOPACOM) have joint responsibilities in the region, and thus, must coordinate force movements. This partitioning of the operational theater makes synchronization difficult and relatively slow. By bringing the Arctic under USEUCOM, which Iceland, Norway, Russia, and half of Greenland are already a part of, the United States can make the region a greater operational priority. This also involves shifting the jurisdiction of Alaska and Canada's northern islands to USEUCOM. USEUCOM would then monitor the bases developed under the new Arctic strategy, while coordinating force movements between these bases, connecting the tactical and strategic levels of this redesign of Arctic military plans.

> **Given Russia's military bases and icebreakers, scholars believe that the West trails 10 years behind Russia militarily in the region.**

On the tactical level, once authority is granted to USEUCOM to carry out a base development strategy, the Secretary of Defense should reassign two Expeditionary Sea Bases (ESBs) from USNORTHCOM to USEUCOM: the USS Lewis B. Puller (ESB 3), which is currently located in Norfolk, Virginia, and the USS Miguel Keith (ESB 5) in San Diego. ESBs are mobile landing platforms that can serve as forward floating bases of maritime security. ESBs 3 and 5 should be relocated and permanently stationed in the Greenland-Svalbard-Norway (GSN) gap, with each one on either side of Svalbard Island. While there are recent talks of Greenland seceding from Denmark, Greenland premier Kim Kielson asserts that the United States-Greenland relationship is "constructive" and experiencing "increased cooperation." The United States. can expect Greenland to support ESBs in their waters regardless of its relationship to Denmark. These bases are not as aggressive as warships but project power through oversight of the two TSR entrances from the European side of the Arctic sea.

## Conclusions

U.S. dominance of the TSR is valuable for economic reasons because it reduces trade time and cost. Yet, in a Mahanian fashion, TSR economic gains are contingent on United States and Russian military competition in the Arctic Circle. Confronted with a trilemma, the United States must pursue two of three policy goals: economic access to the TSR and greater military presence around the route. Cooperation with Russia will only lead to feeble U.S. economic and defense posture in the region. Russia has a clear economic interest in accessing a faster trade route in the TSR and its clear military buildup of the Arctic has forced the U.S. hand to match it with its own military might. The U.S. military can do this by making changes on the strategic, operational, and tactical levels by expanding its number of Arctic bases, shifting jurisdiction of the Arctic to USEUCOM, and placing ESBs in the GSN Gap. The United States. must establish itself as a regional power in the Arctic to realize TSR's economic potential before it opens in the 2040s.

## Print Citations

**CMS**: Georgetown University Center for Security Studies. "The Arctic Trilemma: The United States Must Compete in the Transpolar Sea Route." In *The Reference Shelf: U.S. National Debate Topic 2025–2026: Exploration & Development in the Arctic*, edited by Micah L. Issitt, 105–111. Amenia, NY: Grey House Publishing, 2025.

**MLA**: Georgetown University Center for Security Studies. "The Arctic Trilemma: The United States Must Compete in the Transpolar Sea Route." *The Reference Shelf: U.S. National Debate Topic 2025–2026: Exploration & Development in the Arctic,* edited by Micah L. Issitt, Grey House Publishing, 2025, pp. 105–111.

**APA**: Georgetown University Center for Security Studies. (2025). The Arctic trilemma: The United States must compete in the transpolar sea route. In M. L. Issitt (Ed.), *The reference shelf: U.S. national debate topic 2025–2026: Exploration & development in the Arctic* (pp. 105–111). Grey House Publishing. (Original work published 2024)

# 4
# Climate and Environmental Protection

An Arctic poppy in bloom within the Qausuittuq National Park, Bathurst Island. Photo by Paul Gierszewski, CC BY-SA 4.0, via Wikimedia.

# The Arctic Ecosystem

Does Arctic ice have value? Is there value in preserving the Arctic ecosystems against the destruction from human exploitation and resource extraction? What is the relative value of a polar bear? Estimating the value of the Arctic world is complex, because much of what might be valuable in this landscape is intangible. Ultimately, the fate of the Arctic has always hinged on the balance between conservation and exploitation. Conservation has limited exploitation, but now the exploitation of the world on the broader scale is changing the natural barriers that hampered our capability to exploit the Arctic. As this changes, how will humanity respond? Will the Arctic nations choose to preserve the Arctic for future generations or out of love and respect for this unique wilderness, or will these nations, driven by the hunger for profit, allow Arctic exploitation to further damage and ultimately transform this unique landscape.[1]

Ultimately, this fate depends on the degree to which we place value on the preservation of landscapes for the sake of preserving landscapes. President Theodore Roosevelt, an icon of American conservation, said in his 1908 speech describing "conservation" as a national duty:

> We have become great in a material sense because of the lavish use of our resources, and we have just reason to be proud of our growth. But the time has come to inquire seriously what will happen when our forests are gone, when the coal, the iron, the oil, and the gas are exhausted, when the soils shall have been still further impoverished and washed into the streams, polluting the rivers, denuding the fields, and obstructing navigation. These questions do not relate only to the next century or to the next generation. One distinguishing characteristic of really civilized men is foresight; we have to, as a nation, exercise foresight for this nation in the future; and if we do not exercise that foresight, dark will be the future![2]

The basic scientific knowledge of climate change was also available in Roosevelt's time, and scientists were already predicting that the loss of forests and other habitats coupled with the exploitation of gas and oil, would change the nature of the earth itself. They didn't fully understand what this would mean, but it was imagined that the ices might melt, and the seas rise. Roosevelt was looking at a far more pragmatic type of conservation, preserving entire ecosystems so that we could both preserve the resources needed for industry and also the experience of wilderness for future generations. The question facing Americans as climate change reaches a severe and increasingly dire stage of development, is whether there remains within the nation sufficient interest in the preservation of not only resources but wilderness itself to push back against those who would seek to maximize profit through exploitation and would sacrifice any other consideration in pursuit of this profit. This question will ultimately decide whether Roosevelt's feared "dark future" will be humanity's realized future.

## Historical and Cultural Value

The Arctic is the ancestral home to Indigenous cultures that occupied those regions for thousands of years, descendants ultimately of the first hunter-gatherers who managed to cross the difficult terrain to discover these frozen lands. Preserving the Arctic is therefore, in many ways, key to preserving these traditional cultures.

Indigenous people and groups also play an important role in the effort to conserve the Arctic. This includes leveraging what is called "traditional ecological knowledge," or TEK, to enhance efforts to protect Arctic ecosystems. Conservationist Victoria Qutuuq Buschman explained to the *Wilson Center*,

> The Arctic is the most rapidly warming environment in the world and its Indigenous peoples are experiencing unprecedented change. Our homeland is drastically changing at a rate three-to-four times faster than the global average. The things we see, hear, and smell may cease to exist as we know them before the end of this century, and these dramatic shifts in climatic conditions are impossible to ignore. Across Arctic communities, Inuit and other Indigenous peoples are working to address the pressing climate issues we have inherited in the 21st century, such as: coastal erosion, ocean acidification, pollution, fluctuating and unpredictable wildlife populations, biodiversity loss, unsustainable resource extraction, and data gaps in almost all areas of Arctic research. For us, the effects of a changing climate are tangible and inescapable even as these processes feel like a distant trouble for the rest of the world. We are living, and breathing, change.[3]

For the Indigenous cultures of the Arctic, conservation is also about preserving cultural connections and ways of life that are rapidly disappearing as Arctic resources dwindle. In some cases, Indigenous communities oppose federally mandated and nationally mandated conservation policies, such as those that restrict traditional items of the Arctic diet. Increasingly, Indigenous Arctic people have demanded a voice in making decisions about conservation aims and goals, in an effort to design conservation programs and strategies that are more equitable to the original and ancestral inhabitants of these habitats.

## The Anticonservation Lobby

In April of 2025, Donald Trump signed an executive order prohibiting US states from enforcing climate-oriented policies. Trump said, "These State laws and policies are fundamentally irreconcilable with my Administration's objective to unleash American energy." Trump also suggested that these laws were "putting our coal miners out of business."[4] These and other political actions undertaken by the Trump administration represent the biggest challenge to global conservation and any legitimate effort to combat climate change or to protect the world from its most devastating effects. Trump is not unique in his views, but merely one of the more brazen, outspoken, and authoritarian representatives of a worldview that has always been part of the American debate on the environment; a view that prioritizes profit over preservation, and commerce over conservation.

The problem with the worldview espoused by Trump and like-minded thinkers, according to those who best understand this issue, is that erasing conservation laws and eliminating efforts to reduce petroleum industry use or investments in alternative energy, neither creates jobs or improves the economies of nations in which such actions are taken. Take, for instance, the claim that conservation policies have a negative impact on employment in the coal industry. This claim is fundamentally flawed in many different ways. The decline of the coal industry occurred because natural gas was cheaper to extract, safer, and more profitable for companies. It was the natural evolution of the energy market that caused the decline of the coal industry and so no amount of deregulation will revive the industry. This will not occur until better and more profitable forms of energy are unavailable. Coal may be revived, therefore, but not until energy shortages have reached such a stage that there are no other options.[5] Further, remaining operations to harvest coal have shifted to automated systems, and away from employing miners. Even if the industry is revived, the amount of employment this provides would likely be minimal. Writing for *MIT Sloan Management Review*, Andrew Winston and Hunter Lovins, two capitalists and experts in business innovation, explain that the entire coal industry employs fewer people than the restaurant chain Arby's and argue that trying to revive coal cannot provide meaningful advances in employment, no matter how it is done.[6]

If cancelling environmental regulations won't increase employment and is unlikely to directly benefit consumers, why cancel regulations on things like clean air and clean water? The answer lies in seeing how these regulations impact the companies in question. Environmental regulations and the push for alternative energy don't reduce employment or growth in the fossil fuel sector, but they reduce corporate profitability for the companies involved in fossil fuels. Essentially, coal mine owners will see increased profits. This does not mean they will hire additional workers, but merely that their companies will benefit from being able to more freely pollute and to access resources in areas that had been protected previously. Companies can save money by ignoring measures used to prevent water and air pollution. This is the essential trade-off. When we establish environmental regulations, we cannot do so without forcing fossil fuel companies to adjust in ways that may cost those companies revenues. Disposing of waste in safe ways that do not pollute rivers, for instance, is far more costly than dumping waste into rivers. This cost is exactly why pollution-producing companies spend millions each year to fight against or discourage pollution-control regulations. It took many years to establish the first antipollution laws in the United States, entirely because of the lobby of companies arguing that such regulations were a burden on their businesses because of the extra cost.

Studies have also shown that fossil fuel companies are not direct in their efforts to limit conservation, but purposefully spread misinformation about the connections between conservation and the economy in an effort to reduce popular support for conservation and alternative energy programs. Citizens are being misled to believe that regulation harms the economy, slows growth, and causes fuel prices to increase. Writing for the *Center for American Progress*, Chris Martinez, et al. write, that the $4 trillion fossil fuel industry has also tried to undermine the democratic process, lobbying for, among other things, laws prohibiting environmental activism and protests, while simultaneously "greenwashing," a process by which companies pretend to be more environmentally conscious than can be legitimately demonstrated, thereby clouding the debate over the company's role in the world.[7] First and foremost, however, these companies align themselves with politicians. These politicians promote policies that benefit these companies, and the companies support their political careers and campaigns. A number of politicians have gone on to work in lucrative positions in the fossil fuel industry after their careers in politics, or have come from fossil fuel industry positions into the political sphere.

American citizens are therefore encouraged to see the fossil fuel companies, and the industry, as a key to their own advancement and as benevolent providers of jobs being targeted by environmentalists who care more for nature than the welfare of the average American family. In reality, the choice that Americans make when they support policies like those Trump has used to dismantle environmental controls on petroleum companies, is to prioritize the profit of fossil fuel companies over public welfare, the conservation of resources, and the welfare of future American populations.

## The State of the Arctic

While Donald Trump and many like-minded allies have refuted research on climate change or have suggested that climate change is not as bad as scientists have warned, perhaps even providing benefits, the global scientific studies on climate change is well-established, based on many years of rigorous and dedicated research, and there is no legitimate data to call these findings into question. The world is warming, this warming is due to human activity, and unless humanity dramatically reduces the use of fossil fuels and activities that damage insulating systems (such as the clearing of forests, the storage of waste in landfills, the mass agriculture of cattle, and the production of pollution-producing goods), the environment of Earth will change dramatically and this will have an enormous impact on human societies around the world and will result in the mass extinction of animal and plant life. This is the stark, irrefutable reality of climate change as it is now occurring.

Climate change is occurring everywhere on Earth, but the effects of this change impact different places in different ways. In the Arctic, the pace of warming is accelerated largely because of a phenomenon known as "Arctic amplification," where the warming effects are intensified because the melting of snow ice reduces the albedo effect (reflecting solar radiation back into space), leading to a high degree of heat absorption in the ocean. As a result, the Arctic region has warmed by more than double the global average. What this means, ultimately, is the loss of glaciers and sea ice, and thus the loss of traditional lifestyles, habitats essential for many species, and the rising of the oceans around the world, threatening coastal territories and communities.

In 2023, the National Oceanic and Atmospheric Administration (NOAA), the chief institution for the study of climate in the United States, recorded the warmest summer on record for the Arctic region, along with one of the six warmest years on record. This data came from the NOAA's Arctic Report Card, which documents the progress in combating climate change and the impacts of climate change in the region. After the release of the 2023 Report Card, NOAA administrator Rick Spinrad stated, "The overriding message from this year's report card is that the time for action is now. NOAA and our federal partners have ramped up our support and collaboration with state, tribal and local communities to help build climate resilience. At the same time, we as a nation and global community must dramatically reduce greenhouse gas emissions that are driving these changes."[8] In 2025, Donald Trump announced plans to eliminate climate research at the NOAA and at the National Aeronautics and Space Administration (NASA), the other organization in the United States most involved with producing data to understand climate change.[9]

As the Arctic region continues to disappear, the threats of climate-oriented disaster increase for all the people of the planet. Polar ice caps help to keep the rest of the world cooler. The Arctic is a stabilizing force, a "refrigerator" that cools air currents and stabilizes temperatures across the world. Without this ice, the weather will become more dramatic, and more devastating. First and foremost, the loss of the ice will dramatically increase warming. The impact of climate change, already felt in every part of the plant, will therefore become more extreme. This will mean more extreme winters, and more extreme summers, likely to cause draughts, water shortages, and other disruptions to human societies. Ultimately, this change in temperatures will impact the global food supply. Rising oceanic temperatures will reduce or eliminate entire species of commercial food fish. Crops that once thrives in certain parts of the world will become far more difficult if not impossible to grow.

While these extreme consequences might seem to justify major changes in policy and priority, this does not occur. The primary reason for this is the prioritization of proximate wants over future outcomes. US voters may support a candidate who promises some tangible immediate benefit, like reduced fuel costs or the unlikely availability of mining jobs for Americans hoping that they or their children, might work in a mine. These proximate, tangible advantages to a person's own life take precedence over the idea of preserving or protecting the future of children and unborn generations, outcomes that require a person to have empathy for others or to imagine living the lives of other people. This prioritization of immediate desires also takes precedence over the idea of preserving wildlife, or suggestions that landscapes and ecosystems are resources in themselves, links between living humans and their planet's once lush ecological environment.

Given the advance of the Trump administration's blatant antienvironmentalism and withdrawal of federal support for legitimate scientific research in environmentalism, it remains to be seen whether the United States population has, within it, a sufficient number of genuine conservationists to make meaningful change timely enough to help stem the tide of climate change, in the Arctic or anywhere else on Earth.

## Works Used

Aton, Adam, and Lesley Clark. "Trump Declares War on State Climate Laws." *Politico*, 9 Apr. 2025.

Buschman, Victoria Qutuuq. "Arctic Conservation in the Hands of Indigenous Peoples." *Wilson Quarterly*, Winter 2022.

Luse, Jeff. "Markets Don't Want More Coal: Trump Is Propping Up the Industry Anyway." *Reason*, 10 Apr. 2025, reason.com/2025/04/10/markets-dont-want-more-coal-trump-is-propping-up-the-industry-anyway/.

Martinez, Chris, et al. "These Fossil Fuel Industry Tactics Are Fueling Democratic Backsliding." *Center for American Progress*, 5 Dec. 2023.

"Protecting the Arctic." *Ocean Conservancy*, 2025, oceanconservancy.org/protecting-the-arctic/.

Roosevelt, Theodore. "Conservation as a National Duty." *Voices of Democracy*, 13 May 1908, voicesofdemocracy.umd.edu/theodore-roosevelt-conservation-as-a-national-duty-speech-text/.

Voosen, Paul. "Trump Seeks to End Climate Research at Premier U.S. Climate Agency." *Science*, 11 Apr. 2025, www.science.org/content/article/trump-seeks-end-climate-research-premier-u-s-climate-agency.

"Warmest Arctic Summer on Record Is Evidence of Accelerating Climate Change." *National Oceanic and Atmospheric Administration (NOAA)*, 12 Dec. 2023, www.noaa.gov/news-release/warmest-arctic-summer-on-record-is-evidence-of-accelerating-climate-change.

Winston, Andrew, and Hunter Lovins. "Fossil Fuel Jobs Will Disappear, So Now What?" *MIT Sloan Management Review*, 13 May 2021, sloanreview.mit.edu/article/fossil-fuel-jobs-will-disappear-so-now-what/.

## Notes

1. "Protecting the Arctic," *Ocean Conservancy*.
2. Roosevelt, "Conservation as a National Duty."
3. Buschman, "Arctic Conservation in the Hands of Indigenous Peoples."
4. Aton and Clark, "Trump Declares War on State Climate Laws."
5. Luse, "Markets Don't Want More Coal."
6. Winston and Lovins, "Fossil Fuel Jobs Will Disappear, So Now What?"
7. Martinez, et al., "These Fossil Fuel Industry Tactics Are Fueling Democratic Backsliding."
8. "Warmest Arctic Summer on Record Is Evidence of Accelerating Climate Change," *National Oceanic and Atmospheric Administration (NOAA)*.
9. Voosen, "Trump Seeks to End Climate Research at Premier U.S. Climate Agency."

# We Need Greenland. But Not in the Way Trump Thinks

By Paul Bierman
*Undark*, January 23, 2025

Donald Trump has a thing for Greenland. First, he wanted to buy the Arctic island. Then, his son visited for a photo-op. Now, he refuses to rule out using the U.S. military to seize it.

Decades ago, the value of Greenland was indeed its strategic location between superpowers and its unique mineral resources. No longer. Today, Greenland's value is the ice that covers 80 percent of the island. Keeping Greenland's ice frozen preserves at least a trillion dollars of wealth generation, according to a 2020 study, and would help prevent trillions of dollars in estimated losses by 2100. If we let all that ice melt, global sea level will rise 24 feet higher than it is today. Low-lying farm fields, factories, homes, and large swaths of cities including New Orleans, Miami, Jakarta, and Mumbai will mostly vanish beneath the waves unless we spend even more building barricades to keep the ocean out.

What matters now is cooling Earth's climate to preserve Greenland's ice and with it, our collective coastal future. Yet, the president continues to view the island nation through an antiquated lens of power: possessing the land of others to generate wealth and to keep perceived enemies at bay.

Trump's fascination with Greenland is not a new idea. For more than a thousand years, colonial aspirations have shaped the island's history. The Norse, Danes, Nazis, and Americans came to Greenland chasing territory and natural resources, and for military domination of the Arctic.

Consider Erik the Red, who migrated from Norway as a child, and was later banished from his home in Iceland for murder. In 985 C.E., he sailed deep into a southern Greenland fjord. On a patch of flat coast, Erik established Brattahlíð, the first of what would become about 500 Norse farms on the island. Within a few hundred years, the Norse had vanished. We now know that climate change had pushed sea level almost 10 feet higher in southern Greenland, flooding their settlements and contributing to their demise.

Nearly a thousand years later and a 10-minute boat ride across the fjord from Brattahlíð, U.S. military bulldozers built a mile-long airstrip at the start of the Second World War. Stopping there to refuel on their one-way journey to the European theatre, U.S. bombers rumbled past the Norse ruins.

---

From *Undark*, January 23 © 2025. Reprinted with permission. All rights reserved.

The Americans were not alone. On the desolate East Greenland coast, Nazi meteorologists radioed coded weather reports to Germany, critical information for accurate battlefield forecasts because weather in Europe comes mostly from the west. This was the first hint that Greenland was strategically valuable both in war and as a piece of Earth's climate system.

As the Cold War heated up, the American occupation of Greenland expanded. The sprawling U.S. airfield at Thule, only about 2,800 miles from Moscow, housed atomic weapons and 10,000 men. Camp Century, the Army's nuclear-powered base, hummed inside the ice sheet. It was the start of a grand plan that would have intermediate range tactical nuclear missiles moving back and forth through ice sheet tunnels. This atomic shell game would cover an area about the size of the state of Alabama. But in just a few years, that fever dream faded as the snow tunnels collapsed and long-range ballistic missiles diminished Greenland's deterrent value.

**The last time Earth was as warm as it was in 2024 was likely more than 100,000 years ago—a time when sea level rose at least 20 feet, reshaping the world's coastlines. Some of that water came from Greenland's melted ice.**

By the late 1960s, American soldiers were going home because occupying the Arctic was no longer as strategically meaningful. Abandoning bases, the military left an assortment of hazardous materials in their wake: at Camp Century, 10,000 tons of waste including persistent and toxic organic chemicals like polychlorinated biphenyls, lead, asbestos, and millions of gallons of frozen sewage; and at Thule, a harbor contaminated with plutonium when a nuclear weapon, carried by a crashed B-52 bomber, disintegrated on impact.

In recent years, the Danes have been tidying up that mess, base by abandoned base. But no one has been meaningfully cleaning up the global atmosphere. Since the combustion of fossil fuels started in earnest about 1850, humans have added greenhouse gases sufficient to warm Earth by an average of more than 2 degrees Fahrenheit; the Arctic is warming several times faster.

Like an ice cube on a sultry summer day, Greenland's ice sheet is melting, and that water, flowing into the ocean, is steadily raising sea level globally. The rising ocean is now endangering our national security.

Consider that Naval Station Norfolk, the largest military port in the world and home to five of 12 U.S. aircraft carriers, is now sinking beneath the waves. Worldwide, more than half a billion people live within 25 feet of sea level, according to a 2019 study. If Greenland's ice continues to melt, rising seas will submerge their homes and farms, like those of the Norse. What follows will be the largest human migration in history.

Greenland's ice wouldn't be melting if several decades of moral appeals to address climate change had effectively engaged people and politicians. But those pleadings, often couched in apocalyptic tones, have fallen flat. We need a different approach.

I suggest that advocating for climate stability as a strategic and economic imperative—an idea that politicians across the ideological spectrum could support—is our best hope to keep Greenland's ice frozen and thus the seas from rising further. The U.S. military, integral to the Cold War occupation of Greenland and respected by most Americans, provides an example of just such an approach.

Military and intelligence planners are acutely aware of the strategic threat that climate change presents globally as droughts, floods, and heatwaves destabilize food supplies and regimes. They know that climate change is a threat multiplier and catalyst for conflict. The increasing frequency of rare, extreme, and potentially destabilizing climate events now makes global geopolitics ever more uncertain.

The experience of the U.S. military shows that taking action to reduce climate impacts can be both an economic and strategic win. At bases around the world, engineers have expanded the use of renewable energy while building local microgrids with battery storage for reliable, 24/7 power. Stepping away from fossil fuels is already saving millions of taxpayer dollars in energy costs while reducing supply chain and operational vulnerability.

In a way, Donald Trump has it right. We need Greenland—not the land, but the ice. Keeping that ice frozen means cooling our rapidly warming world, specifically, the Arctic. Decarbonizing the global economy is a critical first step. Next, we need to develop and test technologies, both mechanical and natural, that remove carbon dioxide from the atmosphere. Only then will Earth cool sufficiently to keep Greenland's ice on land and out of the ocean.

The last time Earth was as warm as it was in 2024 was likely more than 100,000 years ago—a time when sea level rose at least 20 feet, reshaping the world's coastlines. Some of that water came from Greenland's melted ice. Even greater melting and sea level rise happened 400,000 years ago. Our planet's temperature, not politicians' ambitions, will determine whether the ice remains on Greenland or melts and pours into the global ocean.

Remember Trump's Florida estate, Mar-a-Lago. Four feet of sea-level rise floods the lower lawn. Six feet and the Atlantic Ocean may trickle into the lobby. Melt Greenland's ice sheet and only Mar-a-Lago's upper story and tower protrude above the waves. Our collective actions over the next decades will determine the future of Earth's ice and thus, global sea level.

President Trump, the most strategic choice now is not buying or seizing Greenland but working globally to save its ice sheet, because what happens in Greenland has yet to stay in Greenland.

**Print Citations**

**CMS**: Bierman, Paul. "We Need Greenland: But Not in the Way Trump Thinks." In *The Reference Shelf: U.S. National Debate Topic 2025–2026: Exploration & Development in the Arctic,* edited by Micah L. Issitt, 122–125. Amenia, NY: Grey House Publishing, 2025.

**MLA**: Bierman, Paul. "We Need Greenland: But Not in the Way Trump Thinks." *The Reference Shelf: U.S. National Debate Topic 2025–2026: Exploration & Development in the Arctic,* edited by Micah L. Issitt, Grey House Publishing, 2025, pp. 122–125.

**APA**: Bierman, P. (2025). We need Greenland: But not in the way Trump thinks. In M. L. Issitt (Ed.), *The reference shelf: U.S. national debate topic 2025–2026: Exploration & development in the Arctic* (pp. 122–125). Grey House Publishing. (Original work published 2025)

# Arctic Has Changed Dramatically in Just a Couple of Decades—2024 Report Card Shows Worrying Trends of Snow, Ice, Wildfire, and More

By Twila A. Moon, Matthew L. Druckenmiller, and Rick Thomas
*The Conversation*, December 10, 2024

The Arctic can feel like a far-off place, disconnected from daily life if you aren't one of the 4 million people who live there. Yet, the changes underway in the Arctic as temperatures rise can profoundly affect lives around the world.

Coastal flooding is worsening in many communities as Arctic glaciers and the Greenland Ice Sheet send meltwater into the oceans. Heat-trapping gases released by Arctic wildfires and thawing tundra mix quickly in the air, adding to human-produced emissions that are warming the globe. Unusual and extreme weather events, pressure on food supplies and intensifying threats from wildfire and related smoke can all be influenced by changes in the Arctic.

In the 2024 Arctic Report Card, released Dec. 10, we brought together 97 scientists from 11 countries, with expertise ranging from wildlife to wildfire and sea ice to snow, to report on the state of the Arctic environment.

They describe the rapid changes they're witnessing across the Arctic, and the consequences for people and wildlife that touch every region of the globe.

## Pace of Change in the Arctic Accelerates

The Arctic of today looks stunningly different from the Arctic of even one to two decades ago. Over the Arctic Report Card's 19 years, we and the many contributing authors to the report have watched the pace of environmental change accelerate and the challenges become more complex.

For the past 15 years, the Arctic snow season has been one to two weeks shorter than it was historically, shifting the timing and character of the seasons.

Shorter snow seasons can challenge plants and animals that depend on regular seasonal changes. Longer snow-free seasons can also reduce water resources from snowmelt earlier in spring or summer and increase the possibility of drought.

The extent of sea ice, an important habitat for many animals, has declined in ways that make today's mostly thin and seasonal sea ice landscape unrecognizable compared with the thicker and more extensive sea ice of decades past.

From *The Conversation*, December 10 © 2024. Reprinted with permission. All rights reserved.

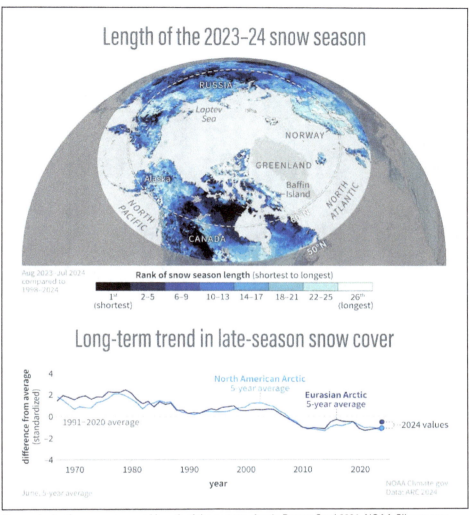

Change in Arctic snow cover and length of the season. Arctic Report Card 2024, NOAA Climate.gov

With a shorter sea ice season, the dark ocean surface is exposed and can absorb and store more heat during summer, which then adds to air and ocean temperature increases. This aligns with observations of long-term warming for Arctic surface ocean waters. Sea ice-dependent animals can also be forced ashore or into longer fasting seasons. The Arctic shipping season is also lengthening, with rapidly increasing shipping traffic each summer.

Overall, 2024 brought the second-warmest temperatures to the Arctic since measurements began in 1900, and the wettest summer on record.

## Arctic Tundra Becomes a Carbon Source

For thousands of years, the Arctic tundra landscape of shrubs and permafrost, or frozen ground, has acted as a carbon dioxide sink, meaning that the landscape

128  Climate and Environmental Protection

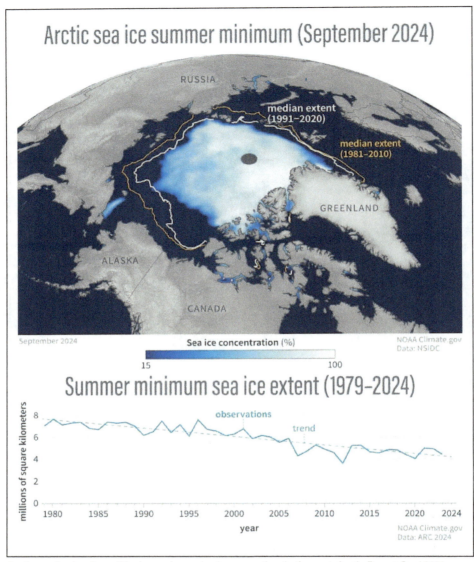

Arctic sea ice has been thinning and covering less area than in the past. Arctic Report Card 2024, NOAA Climate.gov

was taking up and storing this gas that would otherwise trap heat in the atmosphere.

But permafrost across the Arctic has been warming and thawing. Once thawed, microbes in the permafrost can decompose long-stored carbon, breaking it down into carbon dioxide and methane. These heat-trapping gases are then released to the atmosphere, causing more global warming.

Wildfires have also increased in size and intensity, releasing more carbon dioxide into the atmosphere, and the wildfire season has grown longer.

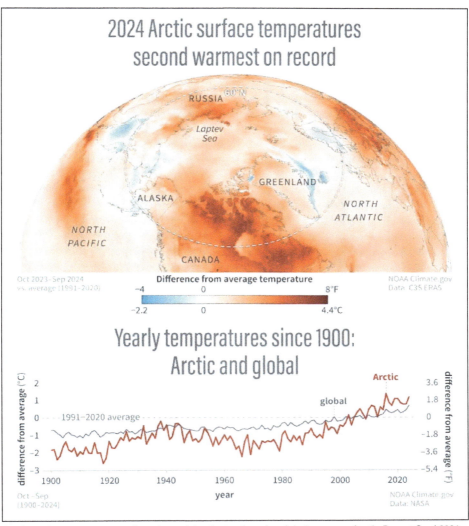

Arctic surface temperatures have been rising faster than the global average. Arctic Report Card 2024, NOAA Climate.gov

These changes have pushed the tundra ecosystem over an edge. Susan Natali and colleagues found that the Arctic tundra region is now a source—not a sink, or storage location—for carbon dioxide. It was already a methane source because of thawing permafrost.

The Arctic landscape's natural ability to help to buffer human heat-trapping gasses is ending, adding to the urgency to reduce human emissions.

## Stark Regional Differences Make Planning Difficult

The Arctic Report Card covers October through September each year, and 2024 was the second-warmest year on record for the Arctic. Yet, the experience for people living in the Arctic can feel like regional or seasonal weather whiplash.

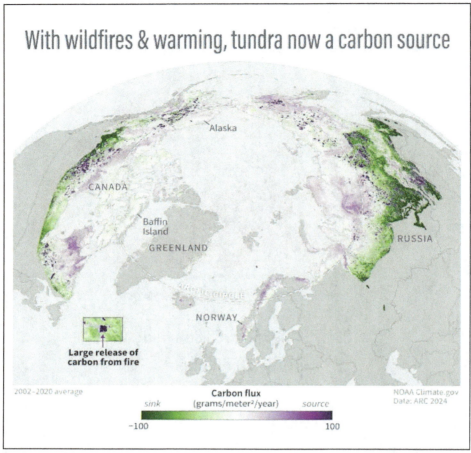

Large parts of the Arctic that once took up more carbon that they released have become net carbon sources in recent years, contributing to climate change. Arctic Report Card 2024, NOAA Climate.gov

Stark regional differences in weather can make planning difficult and challenge familiar seasonal patterns. These include very different conditions in neighboring areas or big changes from one season to another.

For example, some areas across North America and Eurasia experienced more winter snow than usual during the past year. Yet, the Canadian Arctic experienced the shortest snow season in the 26-year record. Early loss of winter snow can strain water resources and may exacerbate dry conditions that can add to fire danger.

Summer across the Arctic was the third warmest ever observed, and areas of Alaska and Canada experienced record daily temperatures during August heat waves. Yet, residents of Greenland's west coast experienced an unusually cool spring and summer. Though the Greenland Ice Sheet continued its 27-year record of ice loss, the loss was less than in many recent years.

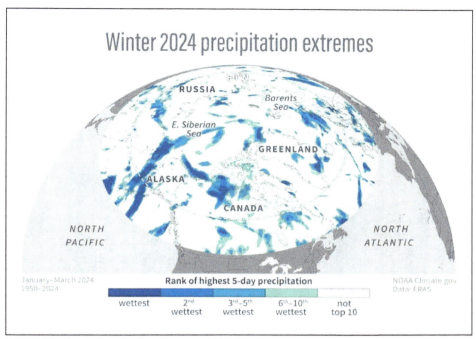

A map of Arctic precipitation from January to March 2024 shows the wide differences among regions and where some saw their wettest year on record. Arctic Report Card 2024, NOAA Climate.gov

## Ice Seals, Caribou, and People Feeling the Change

Rapid Arctic warming also affects wildlife in different ways.

As Lori Quakenbush and colleagues explain in this year's report, Alaska ice seal populations, including ringed, bearded, spotted and ribbon seals, are currently healthy despite sea ice decline and warming ocean waters in their Bering, Chukchi and Beaufort sea habitats.

However, ringed seals are eating more saffron cod rather than the more nutritious Arctic cod. Arctic cod are very sensitive to water temperature. As waters warm, they shift their range northward, becoming less abundant on the continental shelves where the seals feed. So far, negative effects on seal populations and health are not yet apparent.

On land, large inland caribou herds are overwhelmingly in decline. Climate change and human roads and buildings are all having an impact. Some Indigenous communities who have depended on specific herds for millennia are deeply concerned for their future and the impact on their food, culture and the complex and connected living systems of the region. Some smaller coastal herds are doing better.

Indigenous peoples in the Arctic have deep knowledge of their region that has been passed on for thousands of years, allowing them to flourish in what can be an inhospitable region. Today, their observations and knowledge provide vital support for Arctic communities forced to adapt quickly to these and other

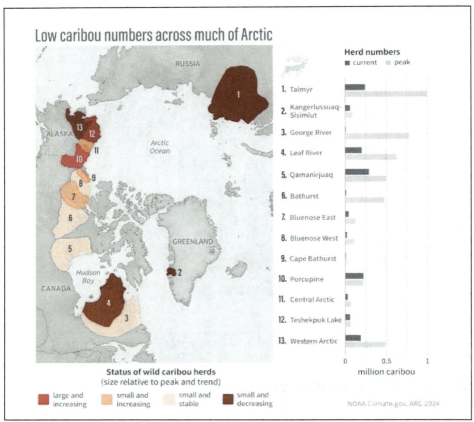

Many caribou herds have lost population in recent years. Arctic Report Card 2024, NOAA Climate.gov

changes. Supporting Indigenous hunters and harvesters is by its very nature an investment in long-term knowledge and stewardship of Arctic places.

## Action for the Arctic and the Globe

Despite global agreements and bold targets, human emissions of heat-trapping gasses are still at record highs. And natural landscapes, like the Arctic tundra, are losing their ability to help reduce emissions.

Simultaneously, the impacts of climate change are growing, increasing Arctic wildfires, affecting buildings and roads as permafrost thaws, and increasing flooding and coastal erosion as sea levels rise. The affects are challenging plants and animals that people depend on.

Our 2024 Arctic Report Card continues to ring the alarm bell, reminding everyone that minimizing future risk—in the Arctic and in all our hometowns—requires cooperation to reduce emissions, adapt to the damage and build resilience for the future. We are in this together.

# Arctic Has Changed Dramatically in Just a Couple of Decades 133

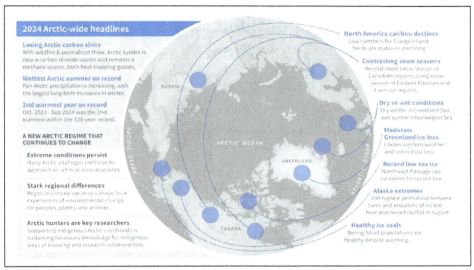

Arctic Report Card 2024, NOAA Climate.gov

## Print Citations

**CMS**: Moon, Twila A., Matthew L. Druckenmiller, and Rick Thomas. "Arctic Has Changed Dramatically in Just a Couple of Decades—2024 Report Card Shows Worrying Trends of Snow, Ice, Wildfire, and More." In *The Reference Shelf: U.S. National Debate Topic 2025–2026: Exploration & Development in the Arctic*, edited by Micah L. Issitt, 126–133. Amenia, NY: Grey House Publishing, 2025.

**MLA**: Moon, Twila A., Matthew L. Druckenmiller, and Rick Thomas. "Arctic Has Changed Dramatically in Just a Couple of Decades—2024 Report Card Shows Worrying Trends of Snow, Ice, Wildfire, and More." *The Reference Shelf: U.S. National Debate Topic 2025–2026: Exploration & Development in the Arctic*, edited by Micah L. Issitt, Grey House Publishing, 2025, pp. 126–133.

**APA**: Moon, T. A., Druckenmiller, M. L., & Thomas, R. (2025). Arctic has changed dramatically in just a couple of decades—2024 report card shows worrying trends of snow, ice, wildfire, and more. In M. L. Issitt (Ed.), *The reference shelf: U.S. national debate topic 2025–2026: Exploration & development in the Arctic* (pp. 126–133). Grey House Publishing. (Original work published 2024)

# Climate Collaborations in the Arctic Are Frozen Amid War

By Jessica McKenzie
*Undark*, April 5, 2022

Earlier in March, temperatures around the North Pole approached the melting point, right around the time of year that Arctic sea ice is usually most extensive. In some places, the Arctic was more than 50 degrees Fahrenheit warmer than average. It's part of an alarming trend; over the past 30 years the region has warmed four times faster than the rest of the globe. The shift is transforming the Arctic land and seascape, causing sea ice to melt, glaciers and ice sheets to retreat, and permafrost to thaw. And while the Arctic is particularly vulnerable to climate change, it also has an outsized potential to contribute to global warming, as melting permafrost releases carbon dioxide and methane into the atmosphere.

And yet, just when the climate scientists and governments across the eight Arctic states should be working together to understand and address the climate crisis, Russia's war on Ukraine has forced the Arctic Council, an intergovernmental group of Arctic states and Arctic Indigenous peoples, to suspend their joint activities in protest of Russia's unprovoked aggression.

"It's so crucial for us to not lose sight of accelerating climate change, and to be making as much progress as possible," Marisol Maddox, a senior arctic analyst at the Polar Institute of the Woodrow Wilson International Center for Scholars in Washington, D.C., said. "And Russia's invasion of Ukraine is creating serious challenges to that within the Arctic context, because of the way that that's manifested with the pause of the Arctic Council, and all of its subsidiary bodies."

As the scientists and experts interviewed for this article all reiterated, the horrific humanitarian crisis on the ground in Ukraine is their foremost concern. But the war is also having all kinds of knock-on effects, from preventing scientists from conducting fieldwork to interrupting long-standing diplomatic relations and research partnerships. This is detrimental to climate science, but not only to climate science.

Jessica McCarty, a professor of geography at Miami University in Ohio and part of the Arctic Council's Arctic Monitoring and Assessment Programm, said scientific and technical exchanges during and after the Cold War helped lay the groundwork for peace by normalizing relationships between countries. "International scientific exchanges are really important for understanding climate change in places like the Arctic, but they're also really important for maintaining things like global peace," she said.

McCarty is currently on sabbatical in Helsinki, Finland, studying how wildfires are impacting the Arctic as part of an E.U.-funded project on black carbon and climate change. But her work on short-lived climate forcers for the Arctic Council is indefinitely on hold, with meetings that had been scheduled in late spring and summer canceled or postponed. She is also barred from doing any work, whether through the Arctic Council or not, with any Russian or Russia-affiliated scientists.

"It sucks," McCarty said about being unable to continue her work for the Arctic Council. "I don't have a good PhD-level word ... except that it sucks."

To the extent that McCarty's work can continue outside the Arctic Council and without the participation of Russian scientists, it will. But the war has been even more disruptive for scientists planning fieldwork within Russia's borders.

Sander Veraverbeke, a professor at Vrije Universiteit Amsterdam, studies climate change and fires in the boreal forest and the Arctic tundra. He is 3 1/2 years into a five-year project studying wildfires in Eastern Siberia, but because of the Covid-19 pandemic and now the war, hasn't been able to conduct fieldwork since 2019.

Siberia is already one of the least-studied parts of the Arctic. "If you look at the amount of data that's available from Siberia, almost for every single variable or field measurement, it's an order of magnitude less, there's much less data," Veraverbeke said. "And the Eurasian Arctic boreal region is about double the size of the North American boreal region. So that already kind of shows that it's really important. Although some of the processes may be comparable across continents, some other processes are not comparable. You cannot just say, 'Oh, we made some measurements in Canada, so in Siberia, it's going to be the same.' For some aspects that may be true, but you can definitely not generalize that."

> "It's so crucial for us to not lose sight of accelerating climate change, and to be making as much progress as possible," said Maddox.

For example, Veraverbeke said the boreal forests in Canada and Alaska are dominated by black spruce, which grow in clumps low to the ground, so fires in those regions burn very intensely. But the forests in Siberia are dominated by *Pinus sylvestris*, a taller pine, interspersed with other tall trees that grow in a more open forest structure, so wildfires will often consume ground fuels but leave the canopies untouched. Part of Veraverbeke's research over the past few years has been quantifying the amount of carbon combusted in these fires, but 2019 was the only year his team was able to get out into the field.

More recently, Veraverbeke secured funding for another project studying how fire impacts permafrost. Does wildfire accelerate permafrost degradation, or can it recover from these events? And what are the long-term effects of fire on greenhouse gas emissions from permafrost? His research is all the more timely because 2019, 2020, and 2021 were all record years for wildfires in Eastern Siberia.

Veraverbeke has tried to make up for the lack of fieldwork by using more satellite data. "Because of that, we've still been able to do some nice science and to be somewhat productive, but it's not the science that we had in mind," he said.

Other researchers studying the geological record of permafrost in Siberia had to cut their fieldwork short after Putin invaded Ukraine and are unsure when they will be able to return. Biologists studying polar bears, salmon, and red knots, a shorebird, have also had to cancel or change their research plans.

The conflict also threatens to disrupt long-standing collaborations with Russian scientists. Veraverbeke's department, for example, has a research collaboration with a branch of the Russian Academy of Science in Yakutsk, which has installed flux towers (a kind of monitoring station) in the forest and tundra to measure greenhouse gases like carbon dioxide and methane, information that is difficult if not impossible to get anywhere else. If war prevents information from being shared between countries for months or even years, it could create significant data gaps in what scientists would otherwise want to be a continuous time series. Of course, climate is not the only scientific field of study impacted.

"It seems like things are only potentially going to be getting worse in Ukraine," Maddox said. "I think we'll be dealing with data gaps from Russia for the foreseeable future."

"It's not just about the access to the research," Maddox added. "Russia is really clamping down on their domestic population. And so there's now heightened concern for Russian researchers, who could, if they're communicating a lot with the West, could be accused of being a foreign agent, because Russia expanded their foreign agent laws recently to not just apply to organizations, but to individuals."

McCarty is also worried about the scientists themselves, in both Russia and Ukraine. "I do have dear colleagues in Russia," she said. "This is where I have to be careful with what I say because I don't want to get them in trouble. I am concerned for them, they are brilliant, and right now that collaboration just cannot continue. Because the conflict and what's happening in Ukraine is paramount, that suffering and that war is what our attention should be on. But there's just a wealth of human capital that's being wasted in Russia."

> "Because the conflict and what's happening in Ukraine is paramount, that suffering and that war is what our attention should be on. But there's just a wealth of human capital that's being wasted in Russia," McCarty said.

Scientists in Ukraine are facing danger and uncertainty of a different magnitude. "I have colleagues in Ukraine," McCarty said. "I've had colleagues who've left Ukraine. And, you know, my concern was with them and their families, initially; it's with them and their futures still, because that is just so uncertain."

Maddox said much of her work right now is focused on how to maintain cooperation between the seven other Arctic States. "There's a lot of discussion about ways to keep this productive cooperation going, even if it doesn't include Russia at this point," she said. "Time is not on our side when it comes to a lot of these issues, and so how can we just keep advancing this and keep the research going and keep making progress on being able to respond if there's some type of environmental disaster or search and rescue incident, for instance."

Climate scientists are hopeful, if not necessarily optimistic, that the conflict will end soon, and they can get back into the field and begin rebuilding relationships with Russian colleagues. Until then, they will make do as best they can with satellite data and ground monitoring elsewhere in the region.

> **Over the past 30 years the region has warmed four times faster than the rest of the globe. The shift is transforming the Arctic land and seascape, causing sea ice to melt, glaciers and ice sheets to retreat, and permafrost to thaw.**

"I just feel sad for international Arctic science right now," McCarty said.

## Print Citations

**CMS**: McKenzie, Jessica. "Climate Collaborations in the Arctic Are Frozen Amid War." In *The Reference Shelf: U.S. National Debate Topic 2025–2026: Exploration & Development in the Arctic,* edited by Micah L. Issitt, 134–137. Amenia, NY: Grey House Publishing, 2025.

**MLA**: McKenzie, Jessica. "Climate Collaborations in the Arctic Are Frozen Amid War." *The Reference Shelf: U.S. National Debate Topic 2025–2026: Exploration & Development in the Arctic,* edited by Micah L. Issitt, Grey House Publishing, 2025, pp. 134–137.

**APA**: McKenzie, J. (2025). Climate collaborations in the Arctic are frozen amid war. In M. L. Issitt (Ed.), *The reference shelf: U.S. national debate topic 2025–2026: Exploration & development in the Arctic* (pp. 134–137). Grey House Publishing. (Original work published 2022)

# Climate Change Is Fuelling Trump's Desire to Tap into Canada's Water and Arctic Resources

By Tricia Stadnyk
*The Conversation*, January 20, 2025

Rising temperatures, a melting Arctic and increasing global water and resource scarcity are behind United States President Donald Trump's threats to make Canada the 51st American state.

A geopolitical storm is brewing in the Arctic, accelerated by climate change and a play for global domination of Arctic land, coastline and trade routes that provide access to massive reserves of critical minerals, oil, gas and water.

The second Trump administration is aware of both the new opportunities and risks as global temperatures shatter new records and thresholds, and an ice-free Arctic becomes a possibility.

## Land Resources

The Arctic is home to a wealth of critical minerals (called rare earth elements) that can withstand extreme cold and pressure changes, making them essential for the space industry, technology and green energy sectors.

China currently controls approximately 60 per cent of the global market, but this past fall announced a ban on exports to the U.S. in retaliation for U.S.-imposed trade restrictions.

Significant reserves of critical minerals exist in Canada and Greenland, explaining Trump's desire to tap into their lucrative resources. And it's not just minerals, but massive reserves of oil and gas, both onshore and offshore, are attracting U.S. interest.

Current political tensions between Greenland and Denmark, with Denmark acknowledging its own neglect, mean the timing couldn't be better for brokering a new U.S.-Greenland partnership.

## Sea Resources

The U.S. Department of Defense tabled a new Arctic Security Strategy last summer in light of increasing security threats associated with the opening of the Northern Sea Route and Northwest Passage.

From *The Conversation*, January 20 © 2025. Reprinted with permission. All rights reserved.

Rights and control over the Northwest Passage remain contested by the European Union and U.S., who rebuke Canada's claim to the route.

If the U.S. claimed the Northwest Passage and the Panama Canal, it would control both major sea routes accessing North America and would represent a significant power grab. Controlling the passage requires a military and strategic presence within the Arctic, which has historically been dominated by Russia.

Canada controls one-third of the Arctic coastline and 25 per cent of the territory, second only to Russia. Climate change is opening passageways—disconnecting land from ice—and creating the potential for land and resource claims that Canada will be forced to defend.

## Water Resources

Canada is perhaps the least prepared to defend threats to water security due to the absence of a national policy or framework over water governance.

Experts have warned against Canada's vulnerability to American interests since at least the 1970s. Trump has already highlighted the "very large faucet" controlling "millions of gallons of water pouring down from the north, with the snow caps and Canada."

He is referring to the Columbia River flowing between British Columbia and Washington and Oregon states, the fourth largest river in the U.S. by volume. The Columbia originates in Canada at the Columbia icefield on the B.C.-Alberta border in the Rockies, but 85 per cent of the basin is in U.S. territory.

The river is governed by a treaty originally ratified in 1964 that resulted in Canada losing control of its rights to divert water for upstream agriculture across the drying Prairie region.

The treaty has been under renegotiation for years to update the terms of water-sharing and controlling interest of both the U.S. and Canada, including new rights for Indigenous Peoples.

**The Arctic is home to a wealth of critical minerals (called rare earth elements) that can withstand extreme cold and pressure changes, making them essential for the space industry, technology and energy sectors.**

An agreement in principle was reached in December 2024 by the Justin Trudeau-Joe Biden administrations, but Trump's second election has sparked concerns due to his expressed desire to divert water to drought-stricken and fire-plagued California. In his inauguration address, Trump falsely claimed the fires in Los Angeles were still burning "without a token defence."

The Canada-U.S. agreement allows both nations to control water within their own jurisdictions to address their greatest needs—energy production and water supply. Need in the U.S. will ways exceed Canada's simply due to the disproportionate populations in each country.

## Canadian Vulnerability

Canada's Transboundary Waters Protection Act deals with water-sharing and is governed by the International Joint Commission's 1909 Boundary Waters Treaty.

And while Canada is the signatory, control and management of water in Canada falls to provincial and territorial governments. In the absence of a national water policy framework, Canada is vulnerable to foreign interests seeking to access, use and control waters crossing the 8,891-kilometre border between Canada and the U.S. in 13 river basins, including those in the Great Lakes.

Canada would be foolish to think this will stop at the Columbia River, particularly as climate change turns up the heat and fires rage across both Canada and the U.S., increasing the desire and competition for a finite supply of water.

Canadians must recognize that climate change is not only changing our landscape and lifestyle, but also exposing imminent threats to our national security and sovereignty. If they value their land and water—and Canadian identity—they must be prepared to respond collectively and decisively to the surging interest over Canada's resource wealth.

### Print Citations

**CMS**: Stadnyk, Tricia. "Climate Change Is Fuelling Trump's Desire to Tap into Canada's Water and Arctic Resources." In *The Reference Shelf: U.S. National Debate Topic 2025–2026: Exploration & Development in the Arctic*, edited by Micah L. Issitt, 138–140. Amenia, NY: Grey House Publishing, 2025.

**MLA**: Stadnyk, Tricia. "Climate Change Is Fuelling Trump's Desire to Tap into Canada's Water and Arctic Resources." *The Reference Shelf: U.S. National Debate Topic 2025–2026: Exploration & Development in the Arctic*, edited by Micah L. Issitt, Grey House Publishing, 2025, pp. 138–140.

**APA**: Stadnyk, T. (2025). Climate change is fuelling Trump's desire to tap into Canada's water and Arctic resources. In M. L. Issitt (Ed.), *The reference shelf: U.S. national debate topic 2025–2026: Exploration & development in the Arctic* (pp. 138–140). Grey House Publishing. (Original work published 2025)

# The Oil Industry's Cynical Gamble on Arctic Drilling

By Rebecca Leber
*Vox*, September 8, 2023

---

The Biden administration can't make a move in the Arctic without a political mess. This week, the administration infuriated the oil industry by canceling seven of the remaining leases in the Arctic National Wildlife Refuge sold by the Trump administration, and proposing new regulations to block oil development in about 40 percent of the National Petroleum Reserve.

Climate activists applauded the decisions. But back in March, Biden raised their ire for approving a vast ConocoPhillips initiative called the Willow Project in the National Petroleum Reserve, which will be unaffected by the new regulations. The sheer size of the Willow Project is at odds with the International Energy Agency's projections that "no new oil and natural gas fields are needed" to make good on the world's net-zero climate promises. It's the largest oil project planned on public lands and will release an additional 9.2 million metric tons of carbon pollution every year, the equivalent of adding roughly 2 million gas-powered cars to the roads.

These fights over the fate of the Arctic seem simple enough: the age-old story of environmentalists versus the oil industry, with the Biden administration caught somewhere in the middle. Yet the reality of what lies behind the oil industry's obsession with this particular part of Alaska is far more complicated.

The Arctic is an especially expensive place to drill for oil, so the price of oil must be high enough to ensure a payoff. Few oil companies in recent years have shown an appetite for taking on that kind of risk, with one major exception: ConocoPhillips. The company's stakes in the Arctic reveal far more than PR statements do about what the oil industry intends. It's essentially a bet that climate action will fail.

## The Arctic Is a High-Risk Environment for the Oil Industry

At the center of the Arctic battle is Alaska's North Slope, which borders the Beaufort Sea in the state's far north. It contains both the National Petroleum Reserve of Alaska (NPRA) and the Arctic National Wildlife Refuge (ANWR). The former drew interest from the private oil developers starting in the Eisenhower administration and the latter held up as a beacon of environmental conservation.

---

From *Vox*, September 8 © 2023. Reprinted with permission. All rights reserved.

Despite what the "National Petroleum" name implies, the area is as prized as ANWR for its ecosystem of beluga whales, walruses, and polar bears, as well as being important to Indigenous communities.

Both areas have been heavily contested ever since. Leaders of the Nuiqsut community, which is about 36 miles from the Willow Project, penned a letter to the Department of Interior this year noting the harm the development would pose to caribou migrations. And ANWR especially, sitting on vast oil reserves, has been a prime target for the industry for decades.

"The Refuge over the years became this marker in the sand for those that wanted to drill," said Kristen Miller, executive director of the Alaska Wilderness League. "If they could get into the refuge, they could get in anywhere."

The industry's lobbying to expand Arctic drilling has spanned every administration since Bill Clinton's. Companies have assumed they would profit from a gusher of oil and from the Alaskan government's oil-friendly, low-taxes position, according to University of Alaska Fairbanks environmental historian Philip Wight. The industry would also benefit from the already-built Trans-Alaska Pipeline, which could already move the oil to the southern port of Valdez for shipment and could avoid an extended fight with environmentalists over building new pipelines.

But these advantages also run up against major barriers that make oil development in the Arctic uniquely difficult—challenges that have far more to do with the environment there than environmental regulations.

The industry aims to squeeze as much as possible out of the cheapest oil reserves it has: areas that will produce a lot of oil for less cost. The Arctic has oil, but it doesn't come cheaply. Companies have to contend with frozen roads, remote areas, and transporting specialized rigs before even unearthing any oil. Even in a world without environmental regulations, it simply costs more for oil companies to drill there, ranking the risks of the Arctic right alongside the risks of deep-water drilling and operating in politically unstable countries. Because of the expense, these are also long-term investments, from which companies plan to benefit over the course of 30 to 40 years. This introduces a lot more uncertainty because of the many factors that can affect oil prices in that time.

The Willow Project faces these disadvantages and more. Willow still faces legal challenges from environmentalists, but the costs of drilling have also gotten worse in other ways—ironically, because of climate change. One example: ConocoPhillips has had to contend with melting permafrost at the sites it intends to drill, which the company will try to neutralize by installing giant chilling devices in the ground.

For Arctic drilling to make sense economically, a company has to bank on prices at the pump remaining high and that consumer demand will still be there for decades to come. That's in spite of expectations that EV sales will cut into demand for gasoline, with EVs on track to become half of global car sales by 2035.

Just to break even, the oil would likely need to sell somewhere between $63 and $84 per barrel, based on an analysis from the World Wildlife Fund higher than what energy analysts expect in a world reducing its reliance on oil.

"They're betting that we're not going to be able to stick within the confines of the Paris agreement," Wight said. "Arctic oil is a fundamental bet on the future and what will and will not happen with the energy transition."

## A Closer Look at ConocoPhillips' Gambit

Given the financial risks, many major players have pulled out of the Arctic region entirely. Royal Dutch Shell has left a door open to still explore in the Arctic but made a splash in 2015 by announcing it would abandon the region, citing the expense of its $7 billion on a failed attempt in the Chukchi Sea between Alaska and Russia. BP sold its holdings in Alaska to the smaller Hilcorp Energy in 2020. Meanwhile, some banks, including JPMorgan Chase, have said they will stop funding loans to oil companies for Arctic development.

Even when the Trump administration offered up ANWR land on a platter with a lease sale late into its term, few companies bothered to show.

**It's the largest oil project planned on public lands and will release an additional 9.2 million metric tons of carbon pollution every year, the equivalent of adding roughly 2 million gas-powered cars to the roads.**

"Basically no major oil companies came to bid at that lease sale," said Miller. "For years we had been saying that this is an area that was too special, too fragile, to develop, but also that it didn't make sense economically. And that's exactly what the results showed." Chevron and Hilcorp have abandoned the ANWR tract they acquired under Trump, entirely voluntarily.

For much of the 2010s, companies had soured on developing expensive oil prospects. Prices have climbed again in the past few years, however, as a result of embargoes on Russian oil and the petering out of shale oil development (and as a global commodity, oil is much more than the Exxons and BPs of the world; 55 percent of global oil is supplied by state-owned oil companies, like in Saudi Arabia and Russia).

"There are some companies now that are making bets again on expensive oil," said Clark Williams-Derry, an energy finance analyst at the nonprofit Institute for Energy Economics and Financial Analysis. "They're basically investing in big capital projects that have a longer lifespan that pencil out when oil prices are higher, $70, $80, or $90 a barrel, but probably wouldn't survive in a world where oil prices can fall to $40 at any moment."

## Oil Companies Are Betting "the World Will Fry"

A company that counts on high oil prices is wagering that climate action will fail. In a world where we meet net-zero targets in the next 25 years, demand for oil and gas will dry up, leaving companies and investors with worthless assets. The industry is intent on that not happening.

The industry also sees the writing on the wall that electric vehicle sales will rise and other demand for its products may slow. But it's counting on demand lingering for decades longer than climate scientists would recommend, even if oil demand does peak in the coming years.

"A peak is not always followed by a collapse," Derry-Williams said. "Sometimes a peak is followed by a bumpy plateau. It's hard to come up with a strong scenario where US gasoline consumption falls dramatically over the next decade or two."

ConocoPhillips may be somewhat unique in the Arctic, but it's not the only company out of alignment with both government pledges and even its own. The major oil companies are all banking on higher oil prices through 2030 than there were from 2015-2020, according to an analysis from Energy Monitor—an expansion strategy, in other words, that depends on global demand to remain very high. They may not be pursuing the Arctic, but they are vying for development where oil and gas are more expensive, like low-quality fracking sites, deep offshore drilling, or politically unstable countries.

"They're basically making the bet the world will fry, and people will continue to buy oil and gas," Derry-Williams said.

**Print Citations**

**CMS**: Leber, Rebecca. "The Oil Industry's Cynical Gamble on Arctic Drilling." In *The Reference Shelf: U.S. National Debate Topic 2025–2026: Exploration & Development in the Arctic*, edited by Micah L. Issitt, 141–144. Amenia, NY: Grey House Publishing, 2025.

**MLA**: Leber, Rebecca. "The Oil Industry's Cynical Gamble on Arctic Drilling." *The Reference Shelf: U.S. National Debate Topic 2025–2026: Exploration & Development in the Arctic*, edited by Micah L. Issitt, Grey House Publishing, 2025, pp. 141–144.

**APA**: Leber, R. (2025). The oil industry's cynical gamble on Arctic drilling. In M. L. Issitt (Ed.), *The reference shelf: U.S. national debate topic 2025–2026: Exploration & development in the Arctic* (pp. 141–144). Grey House Publishing. (Original work published 2023)

# Greenland Is Getting Greener—Helped by a Mining Company and a Group of Tree Enthusiasts

By Adriana Craciun
*The Conversation*, December 6, 2024

As I walked through passport control in Reykjavik en route to Nuuk, the capital of Greenland, the border agent asked me the standard question of why I was travelling. "To look at trees," I answered.

He peered at me suspiciously through the glass and said: "There are no trees in Nuuk."

"Yes there are; people are growing trees there, and there's a forest in south Greenland."

"I think you're lying," he said flatly, not having any of it. The Icelander seemed to be enjoying this little diversion from the rote reasons most travellers give for visiting Greenland: to see icebergs, Viking ruins, or polar bears.

I was indeed coming to Greenland for the plants – my first visit – and the people who cultivate them. That might sound strange, given that 95% of Greenland, the largest island in the world, is covered by an immense ice sheet. Alarming news about Greenland's accelerating ice loss (30 million tonnes per hour) appears daily. Less newsworthy, but equally dramatic, is what is happening on the 5% of Greenland's land that is ice free—it is becoming green.

"Arctic greening" is the climate change phenomenon where land once covered by ice and snow is being colonised by plants. The plants themselves are also changing. Diminutive tundra vegetation is growing taller (so-called "shrubification") and new plants and insects are moving in from the south.

Across the Circumpolar North (a region spanning three continents that includes eight countries), scientists have observed increases in greening of between 20% and 40% in recent decades. The consensus is that greening is accelerating across diverse Arctic regions.

Greening sounds good compared to deforestation, but in the Arctic, the expansion of plant life amplifies dangerous feedback loops. When we think of plants storing carbon from the atmosphere, we typically think of the above-ground plants that we can see. But in the Arctic, carbon storage is mostly below ground, in the frozen soil—permafrost. Permafrost holds more carbon than all the above-ground plants on Earth and twice as much as in the atmosphere.

From *The Conversation*, December 6 © 2024. Reprinted with permission. All rights reserved.

The growth of trees and shrubs accelerates the thawing of permafrost, increasing global heating in a part of the world that is already warming up to four times faster than the rest of the planet. So, Arctic greening doesn't just take over land exposed by retreating ice—it probably accelerates ice melt.

But this kind of Arctic greening is not the phenomenon that I came to investigate. I was interested in the deliberate transformation of Greenland into a literally "green land" via new forests, gardens, and farms. This kind of greening began 1,000 years ago with Erik the Red, the Viking who first farmed Greenland's southern fjords. In one of the earliest real estate boondoggles, Erik named the settlements Greenland (Grœnland in Old Norse) in order to inspire more farmers to follow him to this icy world.

Today, the ancient project of making Greenland green continues in modern, multi-ethnic Greenlandic culture, from mining to gardening to forestry. Intrigued by this convergence of two different kinds of greening—one due to climate change and the other due to human desires—I came to Greenland to understand the connection between them.

So in May 2024 I headed to the Greenland Arboretum in Narsarsuaq, a community of around 100 people near Erik the Red's farms. The arboretum was established in 2004 but its origins in Nordic forestry go back to the 19th century. Nordic foresters have spent decades planting species from northern Asia, Europe and North America, inspired by scientific curiosity into what can grow, and a colonial assurance that they could decide what should grow.

I have been studying the Arctic for many years, from the controversial legacies of Arctic exploration to my forthcoming book on the Svalbard Global Seed Vault, and I was astonished by what I found. Not only is the deliberate cultivation of new plants happening on a remarkable scale, but there exists a near total absence of laws regulating plants—an approach that appears unique to Greenland. In this legal vacuum, it is not only plants that are flourishing. I learned that Greenland's plants are thriving because of the entangled efforts of an international mining company, gardening enthusiasts, farmers, and the dedication of a few remarkable men.

I discovered that significant tree planting is funded by a mining company, keen to show its green credentials by "offsetting" its extraction plans. Amaroq Minerals, which holds the most exploration licenses in Greenland, is a key player in the country's recent critical minerals resource boom. Greenland now rivals China as the planet's main source for rare minerals essential for green technologies.

Extractive industries that embrace such tree-planting schemes may do so because they are keen to show their green credentials by "offsetting" their extraction plans. Amaroq, however, is keen to point out that it "currently has no plans on accounting for any sequestration on carbon offset,"

Though Amaroq was not involved in the initial set up of the arboretum, by funding tree plantations it is joining Greenland's 1,000-year experiment in deliberately greening the land. For decades Greenlanders have been working to introduce exotic plants onto their land "to make the city green," as one gardener told me. It is as if the nation is conducting a large-scale experiment in multiple locations with no controls to see what would happen if the power of plants is unchecked.

## Getting Lost in the Arboretum

The Greenland Arboretum is in Narsarsuaq, a former US base. Today, Hotel Narsarsuaq staff fly Greenland flags out front, reminders of this nation's autonomy in all matters (except foreign policy and security).

I was here to meet with a Danish biologist, Anders Ræbild, who is conducting research alongside the man largely responsible for the aboretum's existence, Kenneth Høegh. Høegh is an agronomist who grew up in Narsaq, a few miles south. He planted his first trees there (Siberian Larch) at the age 14 and he proudly told me they still live today. He now serves as the Head of Representation for Greenland, but in the world of Greenland's trees, he holds a much more influential position.

This was my first day in Greenland, and even before I reached the arboretum's treeline, my ears registered the surrealness of this forest in a supposedly treeless land: birdsong was everywhere. Tiny Common Redpolls were careening through the air, feasting on the seeds of the trees. On this sunny, warm day hovering around 0°C, the forest was ringing with their frenzied pursuit of food and sex. The short Arctic summer was just beginning, and all Arctic life, plants included, takes advantage of these few months with abandon.

I was travelling with my son River, a university student, and we followed the trees as they climbed up the thin soil of a small mountain peak. Neither arboretum nor forest accurately describes this beautiful place. It is too incoherent to be the former but too ordered to be the latter. Stands of 20 or 30-foot Siberian Larches, Norwegian Spruce, Lodgepole Pines from Colorado, Stone Pines from Mongolia, and the occasional deciduous Alaskan Balsam Poplar were arranged together. This is a meeting place of trees from across the northern world, not a coherent forest from any particular part of it.

It is also remarkably large at nearly 400 acres. So large in fact, that though we were supposed to meet in the forest and could hear one another yelling, we couldn't find each another within it. I knew the maths in the abstract but I was still overwhelmed by the numbers on the ground—over 100,000 trees planted in all, from over 100 species, since the 1970s.

## The Man Who Loved Trees

Later that evening, we met Høegh and his colleagues at his cabin for dinner—lamb for them from local farms, cabbage and potatoes for River and me, flown in from Europe like most Greenlandic food.

Høegh is a charismatic man, who was head of the Consultancy Service for Agriculture for the Greenlandic government before he began diplomatic work. After spending several hours in his company, it is clear that regardless of his profession, the trees are his vocation. Høegh described being hand-picked as a student by his predecessor, a Danish forestry professor, along with another forester. The three of them travelled the world for decades, collecting trees. Together these three men appear to have individually directed the planting of the vast majority of trees in Greenland.

At this point, you might be thinking, what about the risks of importing foreign species into an island ecosystem? The potential pests hitching rides in these seedlings? Given that the trees overshadow the tundra plants, what could happen to the local biodiversity given the influx of so many new species? Who is paying for all this? And finally, what are the laws and ethics governing such a large-scale ecological engineering project? These were the questions that brought me to Greenland.

I asked Høegh the question regarding the dangers of importing non-native plants on such a scale. He posed a counter question: when do plants become native? This is a great question, because not just where, but when matters. Plants, like animals, are always on the move and their "native" ranges change over time. We tend to speak of native species in only spatial terms: what is native or non-native to particular places now. When speaking of changes over time, we typically draw the line at what was introduced or eradicated by European colonisation, or humans more broadly.

If people introduce a new plant, this makes it non-native. If animals do it in their fur, feathers, or poop, that is "natural". The assumption is that humans stand apart from nature, and that nature was in pristine balance before humans (especially Europeans) meddled with it. This Edenic idea is prevalent today, in numerous religious and secular beliefs, including conservation sciences. Regardless of where one stands, the consensus that "native" ecologies are inherently good or easily defined is now highly contested.

## Growing a Forest

Høegh's approach to the question of native plants is to take the long view (the extremely long view, in fact) reaching back to Greenland's ancient forested past. Before the last Ice Age, Greenland was forested with the tree species introduced in the arboretum. Most living things in Greenland migrated there after the ice began retreating 10,000 years ago. All are newcomers, with humans only just behind the pioneer plants, arriving 5,000 years ago.

Høegh suggested that since yews, firs, spruce, and poplar were at home in Greenland long ago, why not now and in the future, when our warming climate is expected to resemble that of the Pliocene epoch, between 2 and 5 million years ago.

When I asked him if he is growing a forest, Høegh said modestly that he sees his work "like a stamp collection." He is content that the Arboretum can be different things to different people. For him it is a living collection, and for others, it may be a forest or potentially a source of income or even a sacred place.

Waiting for a ferry in Qaqortoq, I heard a rare local voice of dissent. Fredrik, who grew up there, knew many fellow Inuit who had volunteered to help plant the arboretum. Despite that, he said: "I don't see the meaning of it." The forest does not move him or interest him, he said, and he has no intention of visiting. Fredrik described seeing alarming ecological changes in the last two years: a new kind of snail plagues his yard in large numbers and he is concerned that the trees are bringing in such pests.

## An Experiment without Controls?

I had asked Høegh about the dangers of importing new plant life, given the potential risks of invasive species. He replied that they always obtain a permit to plant. River, an environmental science student, asked how the trees are altering the soil chemistry and affecting established species. The answer was that no one knows. Greenland's Arboretum appears to be a large-scale experiment, but I was starting to wonder: where are the controls?

This is where Greenland's unique political and legal status comes into play. Greenland is one of the few nations that does not recognise international laws against the importation of plants. Laws against the import or export of animals—crucial to hunting and tourism—are very strict. But you can theoretically come into Greenland with any live plant material you like. Once here, there are no clear restrictions on what you can plant in your garden or in open spaces.

I was intrigued by this absence of plant controls. Unlike Denmark, Greenland is not bound by the European and Mediterranean Plant Protection Organization (EPPO) or the International Plant Protection Convention (IPPC). Greenland was exempted from these international treaties governing plant quarantine in 2005.

The head of Greenland's agricultural department confirmed to me that the country "does not currently have legislation regulating plant health". He also directed me to Greenland's 2003 "Law on Nature Conservation" which states that the government "can grant permission" and "can set conditions" for foreign plants.

But when I followed up in person at the Qaqortoq municipality office, where permits would be granted, I was told apparently contradictory information: that planting permits are requested and issued verbally, with no paperwork. There are no forbidden species and there is no scientific input. Another administrator confirmed that there weren't "any standards or rules about what you can bring in or plant." When I asked him about the local appetite for introducing rules, he said these "would be very hard to administer." After speaking to many people in Greenland, it does not appear to me (or to many of them) that plant regulations are applied consistently.

Greenland's exemption from international plant regulations stems from its growing autonomy from Denmark. Colonised by Denmark since 1721, Greenland is a multiracial (Danish and Inuit) and majority-Indigenous autonomous nation moving rapidly toward full independence. After Greenland gained Home Rule in 1979, it pulled out of the European Community (EC) in 1985, and having achieved Self Rule in 2009, it is poised to pursue full independence. Denmark's participation in European and global plant treaties had meant Greenland too followed those rules. Once Greenland left the EC and colonial status, its plants entered a legal limbo.

Greenland's trajectory towards full independence is entwined with its booming extractive industries. Full independence would mean the end of the Danish block grant of DKK 3.9 billion (just over £435 million) that provides about half of Greenland's public budget.

The question now is how Greenland will self-fund its future. After several reversals of policy on uranium and oil exploration, in 2021 a new Indigenous-majority Greenlandic government committed to extracting rare earth elements and minerals for a fossil fuel-free future.

While all eyes have thus focused on Greenland's extractive potential, plants have slipped through the cracks. But my research has uncovered a previously unknown connection between the two.

## Mining, Greening, and Offsetting

Significant tree plantings are currently funded by Amaroq Minerals, which is opening a new gold mine in South Greenland. Registered in Canada, Amaroq is part of a new generation of mining companies extracting the rare earth minerals needed for the wind turbine magnets and car batteries crucial to the green energy transition.

I discovered this link between the trees and mining when I visited the Upernaviarsuk Research Station farm to learn how agriculture is coping with the destabilising climate. I was surprised to see thousands of tree seedlings growing in their vegetable greenhouses. Farm director Kim Neider told me the trees were funded by Amaroq and destined for the arboretum.

In subsequent conversations with Amaroq's leadership, I learned that they have been active in funding Greenlandic tree plantations for several years, and that 2023 was the first year in which they actually produced seedlings, for Qanasiassat, another plantation that Høegh is expanding near the arboretum. In 2023, Amaroq funded the cultivation of 14,500 seedlings, committing DKK 348,500 (worth about £38,740 today). They have agreed to a further $2,000 Canadian dollars per year going forward, although the company has said there is not a limit on this figure.

Amaroq has a social vision almost as ambitious as its extractive vision, fuelled by its extensive exploration licenses and gold mine opening in Nalunaq, South Greenland, in 2024. Using the profits from this mine, its young CEO, Eldur Olafsson, explained to me, they intend to become a "climate company," helping to build a sustainable hydroelectric power grid (currently, most smaller communities run on diesel).

Funding tree planting could give Amaroq carbon credits, which many companies purchase to offset their high carbon emissions—this approach has been criticised by many as a form of greenwashing. Amaroq says it "currently" has no plans to do so—"perhaps in the future," I was told.

But primarily the plantings are a part of their social vision to benefit local livelihoods (like forestry) in the future, Olafsson wrote in an email to me. If this sounds like Høegh's vision for the arboretum, that is because he arranged the funding partnership with Amaroq and calculated their carbon offset, according to Amaroq's executive vice president who confirmed this to me over Zoom and email, sharing the carbon calculation with me. Amaroq later told me this was an "informal calculation."

Granting any carbon offset credit for tree planting in the Arctic is controversial because it generates more greenhouse gases than the trees can capture. One quarter of the northern hemisphere is permafrost, a frozen accumulation of thousands of years of dead plants (and pathogens) storing vast amounts of carbon. The tree roots deepen the active layer of the permafrost (the thin layer of soil that thaws and refreezes each year). They introduce microbial activity into the frozen soil, accelerating the permafrost thaw and allowing for decomposition (and carbon release) to continue through the winter. Scientists describe the smell of thawing permafrost as anything from a "pooey vinaigrette" to a "peaty Scotch." Also, the taller the plants, the more they absorb the sun's energy instead of reflecting it like the snow beneath them would.

Greenland's trees are also capturing carbon dioxide. But because trees grow more slowly in the Arctic than elsewhere, it will take decades before any carbon benefit can balance the carbon they release now. And even at the global scale, in 2023 scientists were alarmed to discover that the "carbon sink" (the plants) we rely on to absorb our carbon emissions failed to absorb a net amount of carbon for the first time ever.

Greenland is pursuing independence and extraction simultaneously in this volatile new climate. Amaroq's involvement with tree planting in Greenland shows that plants are present in both pursuits.

I believe Greenland's laissez-faire attitude to plants is directly related to its move towards independence. It is not an oversight, but a reaction against the policy of isolation that Danish colonisation imposed for so long. For two centuries, Denmark controlled Greenland's contact with the outside world: movement of people, plants, animals and goods to or from Greenland required Danish government permission. This economic and cultural paternalism that Greenlanders endured helps explain their libertarian attitude to plants today. Greenland can at last choose its own laws, and its own plants.

## "We Always Wanted to Live in a Forest"

But Amaroq and the arboretum are not acting alone and are relative newcomers compared to the established efforts of farmers and gardeners when it comes to greening the land. As I visited farms and gardens, I was surprised by the passion of Greenlanders for their plants. But my surprise was due to my own assumptions of what Greenlanders should care about.

When I visited the garden of one Qaqortoq gardener, Alibak Hard, I understood the inadequacy of these assumptions. He told me stories about his individual plants, including Greenland's only maple tree (planted by Høegh) and even a pansy, flowering at my feet in freezing temperatures. I asked Alibak about the source of his passion for exotic plants and he said: "We always wanted to live in a forest."

You can be born and live your life in this Arctic land dominated by ice and yet dream of alien plantscapes—forests full of red-leaved maples, strange flowers underfoot. I told Alibak that many of my fellow academics would say bringing foreign plants into Greenland reflects a colonial, not a Greenlandic, attitude to nature. He replied: "Why shouldn't we change the land like we want?"

Another Qaqortoq gardener told me about the garden club he began in the 1980s, which had over 100 members who would collectively import new plants from Iceland every year. Their goal, he said, was "to make the city green."

When Greenlanders assert such a desire to shape their lands, to diversify the plants in their daily lives, it runs counter to outsider ideals of what Greenland should be. Greenland has been imagined as a sublime icescape onto which outsiders projected their desires—for primordial nature, or authentic Inuit culture unspoiled by modernity. Listening to Greenlanders' views on plants challenged my expectations of what resistance to colonialism looks like. Transforming your lands and plants as you wish can be a powerful exercise of independence.

Ultimately, no one knows how the introduction of hundreds of new species via gardens, farms, and tree plantations is affecting Greenland's ecology now that the warming climate enables these plants to thrive. I collected many anecdotes about ecological novelties—red beetles never seen before, new birds, new snails, an outbreak of caterpillars. The most obvious newcomer required no one to alert me, because this purple plant was everywhere I walked in Qaqortoq: Nootka lupines.

This Alaskan plant was introduced multiple times in Greenland because it can enrich soil by fixing nitrogen and can combat erosion. Gardeners also introduced them for their beauty. Once established, though, Lupines can rapidly become invasive.

Icelandic conservationists recruited lupines in the 1970s to reverse soil erosion, but today those ecosystem engineers have expanded their range by a shocking 35-fold. There is now open biological warfare between locals and lupines in eastern Iceland. Arctic greening appears purple on the eastern slopes of Iceland, and south Greenland may be heading in the same direction.

Without clear laws on which plants cannot be introduced, there is theoretically nothing to stop someone also experimenting with Giant Siberian Knotweed or gorse. Such pioneer plants are brilliant at thriving in disturbed environments and can help establish a succession of new life, even as they decrease biodiversity by crowding out established plants.

Stefano Mancuso is a biologist fascinated by invasive plants because of their remarkable survival skills, which he believes are signs of their intelligence. "The invasive species of today," he writes, "are the native flora of the future, just as the invasive species of the past are a fundamental part of our ecosystem today."

I share the long view with Høegh and Mancuso, that "native" species need to be rethought in both space and time, given the dynamic movements of plants in their 400+ million years evolving on Earth. But in the short term in Greenland, I worry that living solely with this long view without safeguards on moving plants in the present is dangerous.

And this assumption that it is okay for a few men to initiate the transformation of hundreds of acres of communally-owned land is a cavalier one. It resembles the colonial-era attitude that Greenland is a site suitable for social or environmental engineering.

Is it possible to hold these different timescales in mind simultaneously, to act both with short and long-term perspectives? I say yes, it is both possible and necessary to do this. Before releasing thousands of plants onto Greenland's tundra because they can grow there, or they used to grow there, there should be public consultation, considering whether they should grow there today.

> "Arctic greening" is the climate change phenomenon where land once covered by ice and snow is being colonized by plants. Diminutive tundra vegetation is growing taller (so-called "shrubification") and new plants and insects are moving in from the south.

Høegh and his affiliates like Greenland Trees (which accepts international donations to plant trees for carbon offsets) engage local schools to help plant the trees, but it is unclear if ecological education plays a role. Amaroq likewise partners with Siu-Tsiu, the state programme helping young Greenlanders develop new jobs. This may be admirable social engagement, but it follows the lead of select individuals or external groups to green Greenland, rather than a collective desire to do so.

## Future Warming

Greenland had little to fear from pests, wildfires and invasive species in the past due to its cold climate. But today, the risk perception of these hazards lags behind the new warming climate. And Greenland's laws governing plants also lag behind the new climate.

But the embrace of tree planting, cosmopolitan gardening, and farming by many Greenlanders also reflects their extraordinary ability to innovate. Greenlanders are not just skilled at adapting to change, argues anthropologist Mark Nuttall, but at anticipating change proactively.

Extraction and greening are rapidly co-evolving in Greenland, following the expanding networks of mines, roads, military bases, farms and airports. The trees planted by Høegh and Amaroq reveal their seamless integration. I asked scientists if they could foresee what these new forests might look like in 50 or 100 years. They agreed that the situation was too complex to predict.

Yet the scientific consensus holds that "future warming is likely to allow growth of trees and shrubs across much of ice-free Greenland by year 2100," including its far north. The question remains then: will people, from gardeners and farmers to mining corporations, aid more plants in reaching these warming lands, despite the risks?

Greenlanders speak with many voices on questions of what to extract and how, and which plants to grow and where. In this modern society, climate change intensifies the importance of paying attention to the plants in the present, literally and legally. To see the plants in the foreground of a changing Greenland is to perceive the dramatic changes they too are creating at breathtaking speed.

## Print Citations

**CMS**: Craciun, Adriana. "Greenland Is Getting Greener—Helped by a Mining Company and a Group of Tree Enthusiasts." In *The Reference Shelf: U.S. National Debate Topic 2025–2026: Exploration & Development in the Arctic*, edited by Micah L. Issitt, 145–154. Amenia, NY: Grey House Publishing, 2025.

**MLA**: Craciun, Adriana. "Greenland Is Getting Greener—Helped by a Mining Company and a Group of Tree Enthusiasts." *The Reference Shelf: U.S. National Debate Topic 2025–2026: Exploration & Development in the Arctic*, edited by Micah L. Issitt, Grey House Publishing, 2025, pp. 145–154.

**APA**: Craciun, A. (2025). Greenland is getting greener—Helped by a mining company and a group of tree enthusiasts. In M. L. Issitt (Ed.), *The reference shelf: U.S. national debate topic 2025–2026: Exploration & development in the Arctic* (pp. 145–154). Grey House Publishing. (Original work published 2024)

# 5
# Arctic Militarization

U.S. Air Force E-3 Sentry flying above the high Arctic on a 2020 security mission. Photo by U.S. Air Force, via Wikimedia. [Public domain.]

# The Strategic North

What would the world order be like if the United States had won the War of 1812? Had the United States taken full control of Canada, the nation would've had far more stake in the Arctic, but it would also mean that the United States would have undisputed access to the Beaufort Sea. The United States did not, as it turned out, win the War of 1812. It was an unmitigated disaster, leaving Canada an independent state and leaving the United States with only a small chunk of Arctic territory, purchased decades later from Russia, which became the state of Alaska. Because of this history, the United States has limited access to and influence in the Arctic, compared to the nations that have far greater stake in the Arctic region. This doesn't mean, however, that the United States doesn't have both economic and strategic interests in the Arctic region, beyond US ownership of Alaska.

For instance, the United States has interest in what is called the Northwest Passage, a waterway that moves through the Canadian portion of the Arctic. According to the 1825 treaty between Russia and Great Britain—which was later inherited by both the United States and Canada—considers the Northwest Passage to be "internal territory," but the United States has refused to recognize this formulation, and considers the Northwest Passage an international strait, where there is free passage for all nautical traffic. The real underlying issue is access to petroleum deposits in the Beaufort Sea seabed. The United States wants access, but Canada claims ownership. The Trump administration has repeatedly suggested that the United States should take a more aggressive position over Arctic resources, an attitude that is claimed to be motivated by Russian militarization and the increasing importance of, and access too, resources in this part of the world.[1]

The long history of mutual cooperation in the Arctic, between the United States and Canada has meant that this dispute, while still debated, has not intensified to the level of military aggression, but how might things change under the Trump administration, which has caused a rift with the Canadian government and even adopted a hostile posture towards the nation, suggesting that the United States might seek to "acquire" Canada as a new US state, language so aggressive and hostile that the Canadian populace responded with openly anti-American sentiment, including bans on American products. Could the Beaufort Sea become a future battleground between the United States and Canada? Could domestic and economic hostilities, caused by Trump administration actions, intensify US-Canadian competition in the Arctic?

## The History and Nature of Arctic Conflict

The first military disputed involving the Arctic region involved the ownership of important fishing stocks across the region. In 1887 and 1889, the United States seized British-Canadian vessels that were hunting seals in the Bering Sea. This was potentially the first military conflict involving the US naval forces in the Arctic. Ultimately, the conflict proved fruitful, bringing both nations to the table to discuss how best to share resources in the area, and it resulted in the signage of the first international convention on seal hunting. This territory, in the Bering Sea, remains a major source of seafood for many countries around the world and therefore a major source of potential conflict.

While conflicts over resources arose in the late 1800s and into the 1900s, it wasn't until the World Wars that the strategic importance of the Arctic became a theater for direct conflict. The Central Powers (Germany and Austria-Hungary, with the Ottoman Empire and Bulgaria) blockaded the Black Sea and Baltic Sea during the First World War, and this meant that the Allied Powers (Great Britain, France, Russia, and the United States) needed a different route to supply their allies. During the Second World War, which began in 1939, the Arctic was even more essential to the eventual outcome. The trade in things like iron and other minerals, much of which came from Swedish mining, made access to the Arctic Ocean a critical component in the war. Again, the Arctic route was used to ship supplies to Russia, a dangerous passage that saw more than half of many convoys lost to German aircraft and U-boat attacks. Losses in the Arctic were especially high both because the military industries of the world were focused on other theaters of the war, and because the development of the Arctic was minimal.[2]

Another chapter in Arctic warfare can be found in the strategic importance of Greenland to the great powers of the war. Writing for the *Arctic Institute* in 2021, Michael Gjerstad and James Rogers explain that Greenland's importance in the Second World War hinged on the capability to predict the weather that was going to arrive in Europe. As the authors wrote,

> From a meteorological perspective, Greenland is a 'breeding ground' for western Europe's storms. This fact was not lost on the warring nations during the turbulent period between 1939 and 1945, where any weather stations and patrol vessels that could sustain the harsh and unforgiving Arctic would become vital early warning stations for informing military planners of incoming inclement weather.[3]

The lessons learned during World War II, indicating the strategic importance of the Arctic, in multiple ways, and knowledge of how essential the Arctic was for Russia, meant that the region experienced massive militarization during the ensuing Cold War. Intercontinental ballistic missiles (ICBMs), long-range bombers, and other nuclear facilities were constructed across the Arctic, development driven principally by the United States and Russia. Very shortly after this Cold

War began, it was clear that Russia and the United States could not go to war without both nations and much of the rest of the world being destroyed in the process. The militarization in the Arctic was more in the form of a demonstration of influence and power than it was a legitimate building towards conflict. In many ways, this situation never really changed. Though both Russia and the United States focused on other arenas and reduced focus on Arctic security or military capabilities, the strategic importance of the region never really diminished and, as a result, both the United States and Russia continue to have significant investment and military anxieties about the Arctic region.

## Approaching the Modern Arctic

The history of Arctic militarization is important to understanding the current state of affairs in the Arctic and the increase in military tensions since 2014. Writing for the *Arctic Yearbook* in 2019, Rob Huebert explained, political scientist from the University of Calgary explains,

> There is a growing discussion over whether or not the security environment of the Arctic is reentering a "new" Cold War. The crux of the argument is that the era of Arctic exceptionalism is coming to an end. This era has been understood as a period in which the Arctic region was one in which great power rivalries ceased to exist and created an environment in which cooperation and peaceful relations were the core norms. Since the Ukrainian crisis of 2014, there have been growing questions as to whether or not this cooperative environment will be preserved or if the growing tensions between Russia and the West will result in a "new" Cold War in the Arctic. The reality is that there is no new Cold War. Likewise Arctic exceptionalism never really meant the underlying security requirements of the two sides ever really dissipated. Instead what is happening is a renewal of the Cold War with the Arctic as a core location of competition.[4]

While the players are largely the same, the buildup of military forces in the Arctic in the 2010s and 2020s is different than it was in the 1980s. One of the major differences is that the United States is under the control of an administration that is not invested in the North Atlantic Treaty Organization (NATO), long the preeminent body of international mutual military cooperation. NATO is and has always also been an Arctic alliance, as the member states include all those European and North American partners with territories and economic stakes in the Arctic. The shifting perceptions of NATO and US commitments to security partners therefore has direct relevance to the US standing and power in the Arctic.

Donald Trump and allies alleged that NATO has been too dependent on the United States, characterizing this relationship as imbalanced or as European exploitation of US capabilities and military/economic strength. It can be argued that this criticism has some root in truth, as European nations have been able to deemphasize military spending thanks to the US preoccupation with military spending.[5] This occurs, also, because military development is a major factor in the US economy and because the United States reaps vast revenues from a large tax base and spends far less of that revenue on public welfare.

It can be argued, on the contrary, that the balance of influence in NATO and also the balance of spending on defense was purposefully created by the United States in an effort to exert leading influence over the rest of the world. It is this dominant position that gave the United States leverage to position itself as the keystone of Western international governance and the Western economy. The perceived "exploitation" of Europe is therefore, in part, self-serving and has served well to keep the United States in leading positions. This kind of statecraft does not, however, appeal to isolationists and those enthralled with "antiglobalist" sentiments, and the growth in power of this facet of the American populace therefore calls the future of NATO into question.

It is, however, quite straightforward to demonstrate the benefits that the United States still accrues from NATO membership, which is significant. Membership in NATO provides enormous economic benefit that comes from membership in the alliance. IN face all countries that have joined NATO over the years have seen a rise in gross domestic product (GDP) as a result. But what about the United States? Here too membership provides substantive benefits, both reducing the otherwise unilateral costs of intelligence and security and allowing the United States to take advantage of increasing GDPs in member nations.[6]

Like many aspects of the Arctic debate, the advent of climate change is shaping military aims and insecurities involving the Arctic. Writing for the *Arctic Institute* in 2024, Sydney Murkins argues that the melting of sea ice means more energy and resource interest, and that this will exacerbate security concerns in the region. As Murkins writes, "As ice rapidly melts in the Arctic, increased access to critical rare earth elements, such as platinum, copper, lithium, cobalt, and nickel emerge. Moreover, melting ice is opening more easily traversed Arctic waters, which has led to a surge of both military and commercial ship activity in the region. These two developments have increased the potential for conflict in the Arctic."[7]

And meanwhile, as questions about Arctic security and the influence of NATO circulate, experts in international food supplies warn that disputes over fishing and seafood stocks, focused increasingly on the Arctic, are likely to spark new conflicts as well. While most of this conflict will be economic, rather than purely military in nature, conflicts over food supplies could increasingly become incorporated into potential security issues and even military hostility. The increasing number of fisheries, venturing further and further into the Bering Sea, are also the focus of increasing military posturing between NATO states and Russia.[8]

## Managing Conflict

The opening of the Arctic, through the transformation of the climate will have many consequences, some still unforeseen, and is representative of a global change that serves as a harbinger of great loss and suffering; it will also mean new economic and potential security threats. What role the United States will play in this coming age of the Arctic remains to be seen. Will the United States curb Russian ambitions in the region, or will the United States find some way to increase their control of Arctic terrain? These are the unknown variables that may determine how conflict and the fear of conflict continues to shape the Arctic into the next generation.

## Works Used

"Conflict and Geopolitical Issues in the Arctic." *Discovering the Arctic*, 2025, discoveringthearctic.org.uk/arctic-people-resources/resources-from-the-edge/conflicts-geopolitical-issues/.

Evans, Jen, and Andreas Østhagen. "Fisheries Disputes: The Real Potential for Arctic Conflict." *Arctic Institute*, 3 June 2021, www.thearcticinstitute.org/fisheries-disputes-real-potential-arctic-conflict/.

Gjerstad, Michael, and James Rogers. "Knowledge Is Power: Greenland, Great Powers, and Lessons from the Second World War." *Arctic Institute*, 15 June 2021.

Huebert, Rob. "A New Cold War in the Arctic?! The Old One Never Ended!" *Arctic Yearbook*, 2019, arcticyearbook.com/arctic-yearbook/2019/2019-commentaries/325-a-new-cold-war-in-the-arctic-the-old-one-never-ended.

McInnis, Kathleen, et al. "Pulling Their Weight: The Data on NATO Responsibility Sharing." *Center for Strategic and International Studies (CSIS)*, 22 Feb. 2024, www.csis.org/analysis/pulling-their-weight-data-nato-responsibility-sharing.

Murkins, Sydney. "The Future Battlefield Is Melting: An Argument for Why the U.S. Must Adopt a More Proactive Arctic Strategy." *Arctic Institute*, 3 Dec. 2024, www.thearcticinstitute.org/future-battlefield-melting-argument-us-must-adopt-more-proactive-arctic-strategy/.

Rockwell, Keith. "Criticism of NATO Ignores Its Economic Benefit to the US." *Wilson Center*, 29 Mar. 2024, www.wilsoncenter.org/article/criticism-nato-ignores-its-economic-benefit-us.

"What Were the Arctic Convoys?" *Arctic Convoy Museum*, 2025, arcticconvoymuseum.org/history-learning/what-were-the-arctic-convoys/.

## Notes

1. "Conflict and Geopolitical Issues in the Arctic," *Discovering the Arctic*.
2. "What Were the Arctic Convoys?" *Arctic Convoy Museum*.
3. Gjerstad and Rogers, "Knowledge Is Power."
4. Huebert, "A New Cold War in the Arctic?!"

5. McInnis, et al., "Pulling Their Weight."
6. Rockwell, "Criticism of NATO Ignores Its Economic Benefit to the US."
7. Murkins, "The Future Battlefield Is Melting: An Argument for Why the U.S. Must Adopt a More Proactive Arctic Strategy."
8. Evans and Østhagen. "Fisheries Disputes."

# Rising Tensions and Shifting Strategies: The Evolving Dynamics of US Grand Strategy in the Arctic

By Kiel Pechko
*The Arctic Institute*, January 7, 2025

---

The Arctic, often associated with serene landscapes and wildlife, is not just a picturesque region but a strategically important area shaped by environmental, political, economic, and military forces. During the Cold War, it was a battleground between the United States (US), the North Atlantic Treaty Organization (NATO), and the Union of Soviet Socialist Republics (USSR). Post-Cold War, Arctic nations sought to preserve it as a zone of cooperation.[1] However, the strategic competition among the US, Russia, and the People's Republic of China (PRC) has significantly elevated its geopolitical importance in the past decade. The military advancements of Russia and PRC, in contrast to slower efforts by the US and NATO, have turned the Arctic into a critical focus for security, resource access, and control over emerging shipping routes.

In 2009, global interest in the Arctic surged when the US Geological Survey estimated the Arctic held 13 percent of the world's undiscovered oil and 30 percent of the world's natural gas.[2] That same year, the US hosted the Arctic Council Ministerial Meeting to reaffirm cooperation. When Russia asserted its claims over arctic territory, the tone shifted from collaboration to competition. Russia's tenor was not unprecedented; in 2007, Russian submarines planted a Russian flag on the North Pole, proclaiming, "The Arctic is Russian."[3]

The 2013 launch of the US National Strategy for the Arctic Region set the stage for today's geopolitical landscape. By 2022, Russian militarism and PRC's 'Polar Silk Road' ambitions prompted the US to update its Arctic strategy, emphasizing military readiness, infrastructure, and NATO collaboration.[4] The Biden administration further prioritized investments in icebreakers, climate action, and Indigenous involvement, striking a delicate balance between national security and sustainable development in this increasingly contested region.[5]

## Background

Since 2009, the US has issued several Arctic policy documents. The 2009 Arctic Region Policy Directive marked a significant shift in US assessment by recognizing the Arctic's importance, particularly in terms of energy and security.[6] The

---

From *The Arctic Institute*, January 7 © 2025. Reprinted with permission. All rights reserved.

2013 National Strategy for the Arctic Region highlighted energy and US security, environmental preservation, freedom of navigation, access to resources, commerce protection, and peaceful resolution.[7]

The 2022 National Strategy for the Arctic Region was more urgent. Marie-Anne Coninsx, a senior fellow at the Egmont Institute and former EU Ambassador (Arctic), noted that in prior years, US Arctic commitment was "not high on the US Federal Agenda."[8] The Strategy marked a shift in US policy, acknowledging the Arctic as a critical and addressing challenges that had emerged since 2013. It identified four key Pillars: security, climate change/protection, sustainable development, and international governance.[9] The strategy emphasized climate change and investing in sustainable development while increasing capabilities to prevent threats to the US and its allies.[10] Furthermore, it recognized the heightened strategic competition in the Arctic, exacerbated by Russia's invasion of Ukraine.[11] In 2023, US Army General VanHerck advocated to Congress the urgent need for a focused Arctic Strategy, enhanced domain awareness, and the creation of a comprehensive defense framework as these challenges were being driven further by advancing climate change and rising security challenges.[12]

The 2024 Department of Defense (DoD) Arctic Strategy further recognizes the Arctic's growing importance due to climate change, geopolitical shifts, and increasing great-power competition and directs the DoD to "enhance its Arctic capabilities, deepen engagement with Allies and partners, and exercise our forces to build readiness for operations at high latitudes."[13] It highlights critical defense objectives, including protecting US sovereignty, maintaining stability, and ensuring the Arctic's role as a secure avenue for power projection. The 2024 Strategy includes specific initiatives such as enhanced domain awareness; increased communication, intelligence, surveillance, and reconnaissance capabilities; better understanding of cold-weather operations; and conducting key routine training alongside US allies to strengthen joint capabilities.[14] The strategy emphasizes cooperation with, and participation of, Arctic nations and Indigenous communities. It underscores climate adaptation measures. Through these efforts, the DoD seeks to maintain a "monitor-and-respond" posture, ensuring the Arctic remains peaceful while adapting to dynamic environmental and security conditions.

## Climate Change and Security

Climate change has rapidly transformed the Arctic, making previously inaccessible areas more available for economic and strategic use. The North Sea Route (NSR) and the Northwest Passage (NWP) could become economically viable by 2023,[15] potentially diverting traffic from the Suez Canal. As access to critical resources expands, states will shift strategies and intensify competition over the Arctic's growing economic and military significance. This competition has also been driven by its militarization.[16] Over the past decade, Russia has modernized its Arctic military bases, deployed defense missiles, and upgraded its submarine fleet while developing fisheries, petroleum, and mineral extraction.[17] Russia's restrictive NSR claims have further escalated tensions, especially following its invasion of Ukraine. Meanwhile, PRC signaled growing ambitions having ramped up its Arctic investments, researched military applications, and expanded its icebreaker fleet.[18] These developments highlight the Arctic's evolving role in economic opportunity and strategic rivalry, posing significant challenges to US grand strategy.

## American Grand Strategy

The 2009 Arctic Region Policy shifted away from the apathetic post-Cold War US Arctic strategy, recognizing the Arctic as an emerging critical region but only identified two focal points: energy and security.[19] The 2022 National Strategy for the Arctic Region was comprehensive, addressing the Pillars of security, climate change/environmental protection, sustainable development, and international governance. Changes in the physical, financial, and geopolitical landscape have shifted American grand strategy in the Arctic, ending the previous three decades of stability following the Cold War and reflecting rising geopolitical tensions between Russia and Arctic stakeholders.

## Energy

The US is vested in the Arctic's oil and natural gas, with Alaska's Prudhoe Bay Oil Field being North America's most prominent; in the 2000s, Prudhoe Bay typically produced about 8% of America's domestic oil.[20] The importance of Arctic oil to America's energy independence was demonstrated in 2006 when a leak in a critical pipeline at Prudhoe Bay forced the line's shutdown while the US and Iran were in a dispute. At the same time, tensions in the Middle East made it difficult for the US to buy reasonably priced foreign oil. The total cost of oil increased 3% domestically.[21]

The 2009 Arctic Region Policy aimed to increase US oil independence by accessing Alaskan oil deposits. While it received support from energy independence advocates, it was also met with resistance from environmentalists, thus hindering implementation.[22] This debate persisted over the years, shaping subsequent approaches to Arctic resource development. The Trump administration (2021) issued drilling leases on more than 400,000 acres in Alaska's Arctic National Wildlife Refuge.[23] In contrast, the 2022 National Strategy for the Arctic acknowledges the region's economic reliance on hydrocarbon production while advocating for economic diversification, a transition to green energy and environmental protection.[24] In 2024, the US Department of the Interior denied commercial access to a copper deposit through Gates of the Arctic National Park. It banned oil drilling in half of the National Petroleum Reserve-Alaska.[25] Environmentalists and tribal leaders praised these decisions, but Alaska's officials criticized them.[26]

The relationship between environmentalism and energy complicates the US grand strategy in the Arctic, creating inconsistency due to fluctuating politics, public opinion, and stakeholder support. This makes long-term planning for Arctic energy difficult, as leadership changes can shift priorities between resource extraction and environmental protections and alter the trajectory of US Arctic strategy. Such volatility hampers coherent, long-term policies essential for the region's energy security and ecological sustainability, making it a contentious issue influenced by electoral outcomes and shifting public opinion.

## Security

The 2009 US Arctic Region Policy's focus on security has evolved significantly. Previously, the US emphasized unilateral security and preferred bilateral agreements to regional cooperation while ignoring Arctic issues.[27] Post-Cold War, Russia expanded its presence in the Arctic, reverting to Cold War norms. In response, the 2022 National Strategy now prioritizes American people and sovereign territory and rights, committing to enhance capabilities through infrastructure improvements, including an expanded icebreaker fleet, and strengthening cooperation with Arctic allies against Russian aggression.[28]

In his 1987 Murmansk speech, Mikhail Gorbachev described the Arctic as a "peace and cooperation zone," emphasizing collaboration over competition.[29] However during the 2000s, in response to an increased perception of NATO's Arctic presence and strategic importance, Russia began militarizing the region, seeking to reclaim its superpower status after the Soviet Union's collapse.[30] By 2019, Russia had established 14 airfields, six military bases, refurbished 16 Soviet-era deep-water ports, and 10 border posts in the Arctic, with its Northern Fleet boasting an estimated 120 ships, including 40 icebreakers.[31] Russia allowed PRC to construct docks at five of the most significant ports along Russia's Arctic coastline.[32] In contrast, the US needs to develop a centralized command structure in the Arctic, a crucial aspect of its security policy.[33] It has only one significant base at Pituffik Space Base in Greenland, with two additional bases south of the Arctic Circle in Alaska.[34] The US has only two ocean-going icebreakers. There is a stark disparity in military presence between Russia and the US, a potential security dilemma. Russia's action, and the US's inaction, underscore the Arctic's strategic importance and the urgency for a comprehensive security strategy.

Russia has fortified the region to protect future strategic and economic interests.[35] The US has interpreted this as an attempt to achieve regional dominance, prompting numerous military operations in the area.[36] Moscow views US actions as aggressive intrusions into Russian territory, which has led to further militarization on both sides.[37]

This ongoing dynamic was exemplified when US Vice Admiral Lewis claimed that his fleet was prepared to secure sea lanes in the Atlantic and Arctic.[38] A Russian Duma member responded by stating that the US should not tempt fate and that the NSR and Arctic were Russia's responsibility where Russian national interests are present.[39] The current US Grand Strategy has acted firmly but lacks consistent messaging. For example, in 2019, Secretary of State Pompeo decried Russian aggression in the Arctic, citing their regulations on NSR shipping, Arctic fortification, and actions in Ukraine.[40] However, Pompeo's message lacked specificity about why the US felt threatened, simply stating that the US would host military exercises and increase its presence in the Arctic.[41] The US must ensure that its actions in the Arctic are firm, predictable, consistent, and straightforward to avoid further escalating tensions.

The US must improve its messaging on Arctic security by clearly communicating its policies and red lines to Russia to avoid misinterpretations and unintended confrontations. However, it's also important to recognize that the Arctic offers opportunities for cooperation. Enhanced communication can address immediate threats and foster stable regional collaboration. By improving its messaging strategy, the US can more effectively navigate the complex geopolitical landscape and reduce the risks of escalating tensions with Russia.

## Economics

The future of the American Grand Strategy will be influenced by accessibility of the NSR and NWP, which has historically been limited due to short operational seasons (sea ice, weather) and limited infrastructure. These routes are, however, becoming viable with the NSR's navigation season expected to extend from three to six months and the NWP's from two to four months by 2100.[42] If these projections hold, transporting goods from Europe to the Far East via the NSR could be 25 percent more profitable than the Suez Canal Route.[43]

The current shipping industry is already challenged. In 2023, Houthi attacks in the Red Sea led seven major shipping companies to alter their routes, resulting in fewer ships using the Suez Canal. Meanwhile, severe drought in Panama caused a 30 percent decrease in trade since November 2023.[44] These disruptions have led to a 283 percent increase in the cost of a 40-foot container transiting from PRC to northern Europe.[45] These challenges have highlighted the need for alternative shipping routes, making the Arctic routes increasingly important.

The future accessibility of the Arctic makes it more critical than ever however, the West's investment is lagging.[46] In contrast, Russia has prioritized the development of the Arctic to attract traffic, particularly the NSR. From 2011 to 2015, the Russian Arctic doctrine focused on establishing new ports, customs facilities, and marine checkpoints along the Arctic Coastline.[47] Vladimir Putin has stressed the NSR's strategic importance, and Russia's development plan for 2035 includes $19 billion in infrastructure investments.[48] These commitments underscore Russia's recognition of the value of the NSR and the potential increase in cargo transport in the north.

The development of the NSR is also of interest to PRC, who referred to it as part of the Polar Silk Road in its 2018 Arctic White Paper and the 14th Five-Year Plan.[49] Putin's 2023 visit to PRC aimed to enhance the NSR's integration into the Belt and Road Initiative.[50] PRC's NSR use has steadily increased, with transits rising from 27 in 2018 to 62 in 2020.[51] The economic benefits, predicted to be 25 % greater than those of the Suez Canal Route, are also a significant driver of PRC's growing NSR interest.[52] PRC stands to gain significantly from Russian NSR, and Arctic expansion, in general.

In contrast, Europe and Japan have a more complicated view of the NSR. Following Russia's invasion of Ukraine, there was a decline in NSR transit due to sanctions on Russia.[53] However, by 2023, transits had rebounded, with increased transport of Russian LNG to Europe.[54] Despite condemning Russia's actions, Japan has not avoided the NSR, indicating that economic advantages often outweigh political considerations.

Shifting maritime traffic from the Suez Canal to the NSR requires the US to reevaluate trade routes and competitive strategies. As European and Asian partners increasingly use the NSR, they will need to maintain good relations with Russia, which could limit their willingness to oppose Russia, giving Russia a strategic advantage. The US must consider this shift in geopolitical power while also closely monitoring the growing partnership between Russia and the PRC in the Arctic.

> **Increased access to energy and mineral resources extraction due to ice melt, holds significant capacity for growth.**

Increased access to energy and mineral resources extraction due to ice melt, holds significant capacity for growth.[55] The area is home to 22 percent of the world's resources, which includes up to 25 percent of unexplored oil and gas reserves. Despite having no Arctic land, PRC refers to itself as a "near Arctic state," investing heavily in the region. In 2012, PRC invested $12 billion in the Russian Yamal LNG project after US sanctions left Russia short on funds.[56]

Furthermore, in 2018, PRC signed an agreement to invest in Alaskan natural gas export facility,[57] and in 2019, Chinese companies purchased 10 percent ($5.6 billion) of the most prominent Russian natural gas producer.[58] PRC has unsuccessfully attempted to buy mines in Nunavut[59] and recently increased research and land acquisition efforts in the Svalbard.[60] PRC is clearly committed to Arctic resources.

The Arctic's growing accessibility is changing global trade and economics, necessitating an adaptation of US grand strategy. Considering the potential opportunities and challenges posed by Russia's and China's growing commercial influence, the Arctic will play a crucial role in future global trade, resource extraction, and international relations, making it a key consideration for American strategic planning.

## Conclusion

Arctic development and policies have transitioned from cooperative to competitive due to Russian development, militarization, and the PRC's expanding influence, necessitating a recalibrated US grand strategy. The 2022 US National Strategy for the Arctic Region presented a comprehensive response, prioritizing security, climate resilience, sustainable development, and international governance. Its multi-pronged approach, emphasizing the immediate need for a unified NATO Arctic strategy, furthers US interests and aligns military, economic, and environmental policies among Arctic NATO members.

When engaging Russia, the US must present a strong deterrent while pursuing selective diplomatic efforts on shared concerns such as environmental protection and accident prevention. This balanced strategy, valuing both strength and diplomacy, is necessary to reduce the risk of miscalculations. At the same time, strengthening the US role in multilateral forums like the Arctic Council is essential to counter PRC's influence and prevent it from gaining leverage as a self-proclaimed "near-Arctic" stakeholder.

Climate change will create shipping opportunities and natural resource access, accompanied by the potential for economic growth and strategic challenges. In navigating this evolving landscape, the US strives to balance energy, security, and environmental protection but conflicting interests complicate policy coherence. Political changes and stakeholder sentiment can significantly shape strategy, often resulting in abrupt shifts. Maintaining consistency is crucial, especially as environmental issues increasingly intersect with energy independence and national security.

Security remains central to US strategy, but vulnerabilities exist due to military and infrastructure gaps. Russia's aggressive Arctic buildup contrasts with the US's limited fleet and infrastructure. The US relies on strengthening alliances, mainly through NATO, which Sweden now reinforces and Finland's membership. Balancing defense, diplomacy, and coalition-building is vital to safeguarding US interests and Arctic stability. The US can shape the Arctic's future through strategic investments and leadership, ensuring alignment between national security, environmental, and economic goals.

Going forward, the 2024 DoD Arctic Strategy implements essential initiatives of the 2022 Arctic Strategy. It strengthens grand strategy by identifying the Arctic as a critical arena for power projection, alliance strengthening, climate adaptation, and resilience. Its' approach forms a cohesive response to Russian aggression and build-up, PRC ambition, collaboration between Russia and PRC, the addition of Sweden and Finland to NATO and climate change. To that end, it supports improved Arctic capabilities, force readiness for high latitude operations, and deeper engagement with Allies.[61] Given adequate resources and a functioning joined-up government, the DoD strategy will help maintain stability and security in the Arctic.

**References**

1. Congressional Research Service (2024) Changes in the Arctic: Background and issues for Congress. *Congressional Research Service*. 01 October, https://sgp.fas.org/crs/misc/R41153.pdf. Accessed on 1 November 2024
2, 55. Sharma A (2021) China's Polar Silk Road: Implications for the Arctic Region. *Journal of Indo-Pacific Affairs*, 25 October, https://www.airuniversity.af.edu/JIPA/Display/Article/2820750/chinas-polar-silk-road-implications-for-the-arctic-region. Accessed on 1 November 2024

3. Parfitt T (2007) Russia Plants Flag on North Pole Seabed. *The Guardian*, 02 August, https://www.theguardian.com/world/2007/aug/02/russia.arctic. Accessed on 4 November 2024
4, 5, 9, 10, 11, 17, 18, 24, 28. White House (2022) National Strategy for the Arctic Region. *The White House,* October, https://www.whitehouse.gov/wp-content/uploads/2022/10/National-Strategy-for-the-Arctic-Region.pdf. Accessed on 5 November 2024
6, 19, 22, 27. Huebert R (2009) US Arctic Policy: The Reluctant Arctic Power. *University of Calgary, The School of Public Policy*, 2(2): 189–225. Accessed on 31 October 2024
7. White House (2013) 2013 National Strategy for Arctic Region. *The White House.* May, https://obamawhitehouse.archives.gov/sites/default/files/docs/nat_arctic_strategy.pdf. Accessed on 5 November 2024
8. Coninsx M (2022) The New US Arctic Strategy. Welcome Back, America! *Egmont Royal Institute for International Relations*, 15 December, https://www.egmontinstitute.be/the-new-us-arctic-strategy-welcome-back-america. Accessed on 5 November 2024
12. Roza D (2023) NORAD Boss asks Congress for Better Domain Awareness. *Air & Space Forces Magazine*, 23 March, https://www.airandspaceforces.com/norad-radar-china-balloon-congress. Accessed on 4 November 2024
13, 14, 61. US Department of Defense (2024) 2024 Department of Defense Arctic Strategy. *Department of Defense*, June, https://media.defense.gov/2024/Jul/22/2003507411/-1/-1/0/DOD-ARCTIC-STRATEGY-2024.PDF. Accessed on 19 December
15, 16. Pincus R (2020) Three-way power dynamics in the Arctic. *Strategic Studies Quarterly*, 14(1): 40-63, Spring, https://www.jstor.org/stable/26891883. Accessed on 4 November 2024
20, 21. Clifford K and Peters JW (2006) Lessons from Prudoe Bay. *New York Times*, 09 August, https://www.nytimes.com/2006/08/09/opinion/09wed2.html. Accessed on 4 November 2024
23. Groom N (2021) Trump Administration Issues Last-Minute Arctic Refuge Drilling Leases. *Reuters*, 21 January, https://www.reuters.com/article/world/us/trump-administration-issues-last-minute-arctic-refuge-drilling-leases-idUSKBN29O2K4/. Accessed on 5 November 2024
25, 26. Friedman O (2024) Biden Shields Millions of Acres of Alaskan Wilderness from Drilling and Mining. *The New York Times*, 19 April, https://www.nytimes.com/2024/04/19/climate/biden-alaska-drilling-mining-nrpa.html. Accessed on 4 November 2024
29, 30. Boulègue M (2019) Russia's Military Posture in the Arctic Managing Hard Power in a 'Low Tension' Environment. *Chatham House*, 28 June, https://www.chathamhouse.org/sites/default/files/2019-06-28-Russia-Military-Arctic.pdf. Accessed on 5 November 2024

31, 34. Postler A (2019) Contextualizing Russia's Arctic Militarization. *Georgetown Security Studies Review*, 18 February, https://georgetownsecuritystudiesreview.org/2019/02/18/contextualizing-russias-arctic-militarization/. Accessed on 4 November 2024

32, 33. Filijovic M and Jardine S (2024) Russia's Queenside Castling in the High North: A Strategic Risk or Opportunity for the West? *The Arctic Institute*, 08 October, https://www.thearcticinstitute.org/russias-queenside-castling-high-north-strategic-risk-opportunity-west/. Accessed on 4 November 2024

35, 36, 37, 38, 39, 40, 41. Peterson MB and Pincus R (2021) Arctic Militarization and Russian Military Theory. *Orbis*, 65(3): 490-512. Accessed on 3 November 2024

42, 43, 52. Khon VC et al (2010) Perspectives of Northern Sea Route and Northwest Passage in the Twenty-First Century. *Climatic Change*, 100(3–4): 757–768, June, Accessed on 2 November 2024

44, 45. The Economist (2024) How Viable is Arctic Shipping. *The Economist*, 18 January, https://www.economist.com/the-economist-explains/2024/01/18/how-viable-is-arctic-shipping. Accessed on 4 November 2024

46. Bennett M (2016) Is the Northwest Passage Too Small to Compete with the North Sea. *Anchorage Daily*, 29 June, https://www.adn.com/commentary/article/northwest-passage-too-small-compete-northern-sea/2011/08/25/. Accessed on 5 November 2024

47. Blunden M (2012) Geopolitics and the Northern Sea Route. *International Affairs*, 88(1):115–129, 20 January. Accessed on 1 November 2024

48. TASS Arctic Today (2023) Expert: Northern Sea Route May Become Russia's Biggest Revenue Source in Arctic. *TASS*, 25 October, https://tass.com/economy/1696643. Accessed on 5 November 2024

49, 51. Martins TT (2023) Arctic Ambitions: China's Engagement with the Northern Sea Route. *The Diplomat*, 24 November, https://thediplomat.com/2023/11/arctic-ambitions-chinas-engagement-with-the-northern-sea-route. Accessed on 05 November 2024

50. Chen L and Soldatkin V (2023) Putin Praises Dear Friend Xi, Pitches Russia's North Sea Route. *Reuters*, 18 October, https://www.reuters.com/world/putin-praises-xi-pitches-russias-northern-sea-route-2023-10-18/. Accessed on 5 November 2024

53, 54. Hilde PS et al (2024) Cold Winds in the North: Three Perspectives on the Impact of Russia's War in Ukraine on Security and International Relations in the Arctic. *Polar Science* 41: 101050, September. Accessed on 3 November 2024

56, 57. Dillow C (2018) Russia and China Vie to Beat the US in the Trillion-Dollar Race to Control the Arctic. *CNBC-Global Investing*, 06 February, https://www.cnbc.com/2018/02/06/russia-and-china-battle-us-in-race-to-control-arctic.html. Accessed on 5 November 2024

58. Humpert M (2022) Chinese Shipping Company Cosco to Send Record Number of Ships Through Arctic. *High North News*, 13 September, https://www.highnorthnews.com/en/chinese-shipping-company-cosco-send-record-number-ships-through-arctic. Accessed on 5 November 2024
59. Strong W (2020) Ottawa blocks Chinese Takeover of Nunavut Gold Mine Project After National Security Review. *CBC News*, 22 December, https://www.cbc.ca/news/canada/north/canada-china-tmac-1.5851305. Accessed on 5 November 2024
60. Agence France-Press (2024) Norway Blocks Sale of Last Private Land on Svalbard After Chinese Interest. *The Guardian*, https://www.theguardian.com/world/article/2024/jul/01/norway-blocks-sale-last-private-land-svalbard-china-interest. Accessed on 5 November 2024

**Print Citations**

**CMS**: Pechko, Kiel. "Rising Tensions and Shifting Strategies: The Evolving Dynamics of US Grand Strategy in the Arctic." In *The Reference Shelf: U.S. National Debate Topic 2025–2026: Exploration & Development in the Arctic*, edited by Micah L. Issitt, 163–172. Amenia, NY: Grey House Publishing, 2025.

**MLA**: Pechko, Kiel. "Rising Tensions and Shifting Strategies: The Evolving Dynamics of US Grand Strategy in the Arctic." *The Reference Shelf: U.S. National Debate Topic 2025–2026: Exploration & Development in the Arctic,* edited by Micah L. Issitt, Grey House Publishing, 2025, pp. 163–172.

**APA**: Pechko, K. (2025). Rising tensions and shifting strategies: The evolving dynamics of US grand strategy in the Arctic. In M. L. Issitt (Ed.), *The reference shelf: U.S. national debate topic 2025–2026: Exploration & development in the Arctic* (pp. 163–172). Grey House Publishing. (Original work published 2025)

# India's Arctic Challenge: Aligning Strategic Interests with Regional Realities

By Nima Khorrami
*The Diplomat,* September 30, 2024

India's Arctic strategy remains in its formative stages, reflecting the relatively undeveloped nature of the country's Arctic policy. Despite the limited body of literature on the subject, however, a recurring theme advocates for greater Indian engagement in the Arctic to better safeguard and secure India's expanding interests both in the region and globally.

India's interests in the Arctic are primarily centered on scientific cooperation, environmental monitoring, and, more recently, resource security. For instance, the Arctic holds significant importance for Indian scientists due to its direct impact on India's monsoon patterns, which are crucial for the agricultural sector and the country's overall food security. Additionally, the increasing accessibility of the Arctic presents New Delhi with opportunities to meet its growing energy demands and address its shortage of strategic rare earth elements (REEs)—notwithstanding its awareness that the harsh environmental conditions, high costs of extraction, and the need for advanced technology pose substantial barriers.

A notable, and relatively recent, development in India's Arctic strategy is the emerging narrative among Indian strategists that posits New Delhi's engagement with Russia in the Arctic as a strategic necessity to counter the perceived rise of Chinese influence in the region. This discourse seeks to justify India's push for a greater regional presence, and its close cooperation with Russia, as a means to counter Beijing's growing interests in the Arctic and its broader influence over Moscow.

India is being portrayed as a counterbalance to China, offering Russia a viable alternative to avoid international isolation and over-dependency on Beijing. Given the perceived importance of the Russian Arctic in addressing India's energy and mineral security and growing concerns with regard to a China-Russia partnership in the Arctic, enhanced bilateral cooperation between India and Russia in the Arctic is held up as a welcomed alternative to China's influence; one that would satisfy both India's and its Western partners' strategic interests.

In an important sense, this approach is the mirror image of New Delhi's strategy in Central Asia, where it has sought to counter China's expanding influence

From *The Diplomat*, September 30 © 2024. Reprinted with permission. All rights reserved.

—viewed as a strategic vulnerability—by, amongst other things, leveraging Russian concerns about China's growing presence. The argument suggests that India's strategic interest lies in diversifying Russia's partnerships away from China by providing alternatives in markets, finance, and manpower.

However, much like its strategy in Central Asia, New Delhi's approach to the Arctic may be unsuccessful for similar reasons. Its strategy appears to be a reactionary stance focused more on countering China than on responding to the specific dynamics of the Arctic region. By emphasizing its close relations with the Arctic 7 (A7) to justify its reliability in working with Russia, India overlooks a critical point: While most A7 countries view China's growing influence in the Arctic as problematic, they do not consider it an immediate threat to regional cooperation. Instead, it is Russia—the very actor to which India seems eager to anchor its regional policy—that raises the most concerns.

This tension is most evident in the calls to support Russia's proposal for greater BRICS+ engagement in the Arctic. The reasoning behind this is that coordination between India and Russia within BRICS is crucial to avoid surprises from joint Russian and Chinese proposals and initiatives in the Arctic. However, India's Western partners are likely to reject such calls as they promote greater multipolarity in Arctic governance—an idea that none of the Arctic states, including Russia until recently, has favored. Arctic states prefer to maintain the current exclusive arrangements in regional governance, where only states with Arctic territory are entitled to participate in and shape the regional governance structure.

**The Arctic holds significant importance for Indian scientists due to its direct impact on India's monsoon patterns, which are crucial for the agricultural sector and the country's overall food security.**

India's support for the Russian proposal, while serving its interests in limiting Russia-China cooperation and Russia's desire to avoid over-reliance on Beijing, stands in stark contrast to its Western allies' preference for exclusivity. As a result, it is unlikely to be welcomed in Western Arctic capitals.

Instead of aligning with Russia, India should consider forging an approach with other like-minded non-regional states such as Japan and South Korea, both of which share India's concerns about the growing China-Russia cooperation in the Arctic and the potential for their businesses to lose out on Arctic opportunities to Chinese competitors. The trio should then advocate within the Arctic Council for enhanced observer participation to rectify current disparities and promote a more inclusive and equitable Council, fostering a balanced Arctic perspective.

To this end, the underlying goal must be twofold. First, to convince the Nordic states, as well as the United States and Canada, to more rapidly open up their Arctic regions to commercial activities thereby providing a real alternative to Russia's Arctic, where most commercial activities are currently concentrated.

Second, to push for a reformed Arctic Council—one with the mandate to debate hard security issues—to avoid polarization and fragmentation in the regional governance system whereby Russia and its partners adhere to one set of rules while the A7 states follow another.

## Print Citations

**CMS**: Khorrami, Nima. "India's Arctic Challenge: Aligning Strategic Interests with Regional Realities." In *The Reference Shelf: U.S. National Debate Topic 2025–2026: Exploration & Development in the Arctic,* edited by Micah L. Issitt, 173–175. Amenia, NY: Grey House Publishing, 2025.

**MLA**: Khorrami, Nima. "India's Arctic Challenge: Aligning Strategic Interests with Regional Realities." *The Reference Shelf: U.S. National Debate Topic 2025–2026: Exploration & Development in the Arctic,* edited by Micah L. Issitt, Grey House Publishing, 2025, pp. 173–175.

**APA**: Khorrami, N. (2025). India's Arctic challenge: Aligning strategic interests with regional realities. In M. L. Issitt (Ed.), *The reference shelf: U.S. national debate topic 2025–2026: Exploration & development in the Arctic* (pp. 173–175). Grey House Publishing. (Original work published 2024)

# Resource Wars: How Climate Change Is Fueling Militarization of the Arctic

By Joanna Rozpedowski
*RealClear Defense*, August 7, 2024

As the Russo-Ukrainian conflict continues to redefine the international security landscape, more sources of dispute among rivals are emerging. New outbreaks of mass violence in the Middle East and Africa, struggles over drinking water from Afghanistan to Niger, ethnic cleansings, simmering conflicts, unrests, insurgencies, and civil wars in the earth's poorest countries, and endless flows of migrants and refugees into Europe and the United States are only a few yet concerted and unabating forces reorganizing the geo-economic, geopolitical, and social landscape which deepen social cleavages and increase prospects for conflict.

The appreciation of the multifaceted threat vectors arising from localized environmental degradation and resource and food scarcity presents one of the more compelling challenges for policymakers. In his 2012 book *Climate Wars: What People Will be Killed in the 21st Century,* Harald Welzer argued that the consequences stemming from the relationship between violence and climate change will "establish different social conditions from those we have known until now" which will "spell the end of the Enlightenment and its conception of freedom" resulting in the decline of the Western social model based on democracy and liberalism. Resource scarcity and relative economic deprivation will lead to violence, and violence, Welzer contends, is always highly adaptive and "innovative" as it develops "new forms and new conditions" for the use of force. The twentieth-century wars may well have been driven by conflicts over land, religion, and economic interests but the wars of the 21st century will in no insignificant measure be fueled by the multipronged environmental crises leading to a potential state failure contagion, water and food shortages, and a brutal scramble for resource dominance in strategic areas of the globe. The world's rapacious appetite for natural resources and rare earth minerals will open new theatres of potential conflict ranging from outer space to the Arctic.

## The Arctic's Military Landscape

Melting ice in the Arctic is unlocking a valuable natural resource base, oil and gas deposits, and minerals necessary for fueling "green" economic growth already pushing countries to jostle over territorial claims and control of the region's ship-

From *RealClear Defense*, August 7 © 2024. Reprinted with permission. All rights reserved.

ping routes and navigation rights. The world's insatiable hunger for the vast Arctic deposits of oil, gas, nickel, copper, lead, zinc, diamonds, gold, silver, manganese, titanium, and abundant fisheries worth trillions of dollars present a substantial incentive for cooperation and an even stronger propensity toward conflict.

The militarization of the Arctic is escalating, with contiguous and non-contiguous states attempting to secure their claims through the reopening of Cold War-era military sites, airfields, nuclear and submarine facilities, deep-water ports, military exercises, and testing of logistically relevant assets and new weapon systems. Since 2005, Russia has significantly modernized its navy and invested in hypersonic missiles capable of evading U.S. sensors and compromising undersea cables and communication infrastructure. It is also developing and leveraging its icebreaker fleet to ensure year-round navigability of the Northern Sea Route, which runs along Russia's Arctic coast and offers reduced transit times between Europe and Asia.

The Arctic holds significant strategic value for China—a self-declared "near-Arctic" nation, as it offers access to new shipping routes and vast resources. China, too, harbors strategic ambitions as it attempts to incorporate the Polar Silk Road into its wider Belt and Road Initiative and develop Arctic trade routes.

Through Sino-Russian cooperation initiatives, China and Russia are keen on securing maritime power and economic interests in the High North, while the NATO pact members seek to cement their strategic presence *qua* influence through NATO's Arctic Command, a deterrent coordination and cooperation measure aimed at consolidating and preparing joint naval assets for the requisite force projection and prospective combat operations in challenging polar conditions. NATO's Regional Plan North aims to equip the Alliance with adaptive capabilities that anticipate challenges, protect the freedom of navigation, and equip its members with interoperable surveillance and reconnaissance capabilities. After Sweden and Finland joined the Alliance in 2024, NATO is well placed to enhance its deterrence posture and surveillance capabilities in the region as seven out of eight of the Arctic countries are now members.

**The wars of the 21st century will in no insignificant measure be fueled by the multipronged environmental crises leading to a potential state failure contagion, water and food shortages, and a brutal scramble for resource dominance in strategic areas of the globe.**

The United States is investing heavily in new air stations and naval base installations above the Arctic Circle through the Supplementary Defense Cooperation Agreement signed with Norway in April 2021 all in the name, as the U.S. 2022 Arctic Strategy contends, of ensuring the region remains "peaceful, stable, prosperous, and cooperative." To maintain competitive advantage in the region, the U.S. military has expanded its presence in the Arctic by establishing Alaskan Command and increasing investments in icebreakers, surveillance systems, and cold-weather training. U.S. outer space assets prove consequential in this endeavor as they provide the necessary space-situational awareness capabilities that inform operational readiness and report on weather conditions. Yet, according to the International Institute for Strategic Studies (IISS), the balance of power remains at present tilted in favor of Russia.

According to experts, increasing cooperation between China and Russia does not necessarily signify the formation of an anti-Western alignment and a "coordinated revisionist strategy" in the Arctic. Western states, however, must address the challenges posed by Arctic near-peer competitors bent on undermining their assets and adapt to the changing Arctic geopolitical landscape as tensions in the region rise and threaten to spill over into yet another potentially perilous armed confrontation.

While China has strategically refrained from openly articulating its military interests in the Arctic, it nevertheless aims to position itself as a global power with considerable maritime strength capable of exerting influence over the region's norms of engagement. Its investments in building its navy and leadership in resource extraction, scientific research, commercial fishing, shipping, and maritime infrastructure aim to consolidate a viable foundation for its Arctic Silk Road and thus integrate the region into Beijing's broader geopolitical and military strategy. Those do not inherently contradict Russia's aspirations but neither do they complement them. Moscow is weary of China's potential dominance in the region; however, in the name of pragmatism, both sides are eager to benefit from their mutual yet complex economic relationship and potential for joint investments and resource development.

Diplomatic frictions will likely pave the way for pronounced regional and international instabilities—or resource wars—between major Arctic powers and their allies. The victors in the scramble for the Arctic, or the race to become the next great Polar power, will have the authority to unilaterally define territorial boundaries and establish new rules of engagement in the region. Time will tell if the Arctic Council will play a significant role in promoting cooperation and mitigating the threat of overt militarization of the region. With significant governance gaps in its mandate, which explicitly excludes military security matters, and legal ambiguities within international treaties and agreements, the Arctic security dynamics will undoubtedly become a subject of considerable contention in the coming years.

## Print Citations

**CMS**: Rozpedowski, Joanna. "Resource Wars: How Climate Change Is Fueling Militarization of the Arctic." In *The Reference Shelf: U.S. National Debate Topic 2025–2026: Exploration & Development in the Arctic*, edited by Micah L. Issitt, 176–179. Amenia, NY: Grey House Publishing, 2025.

**MLA**: Rozpedowski, Joanna. "Resource Wars: How Climate Change Is Fueling Militarization of the Arctic." *The Reference Shelf: U.S. National Debate Topic 2025–2026: Exploration & Development in the Arctic*, edited by Micah L. Issitt, Grey House Publishing, 2025, pp. 176–179.

**APA**: Rozpedowski, J. (2025). Resource wars: How climate change is fueling militarization of the Arctic." M. L. Issitt (Ed.), *The reference shelf: U.S. national debate topic 2025–2026: Exploration & development in the Arctic* (pp. 176–179). Grey House Publishing. (Original work published 2024)

# Why the US Is Losing the Race for the Arctic and What to Do About It

By Josh Caldon
*Center for International Maritime Security (CIMSEC)*, April 13, 2023

Almost weekly there is another story insinuating that the US is losing the "race for the Arctic." Those who support the claim that the US is losing this race often highlight that the Arctic ice is melting and that this environmental change is opening up potential trade routes and making natural resources more ripe for exploitation. Others then point out that Russia has increasingly re-militarized the Arctic and that China has also made inroads to establish itself in the region.

One key point these articles often make is the United States' relative lack of icebreakers compared to its competitors. What is missing from this conversation, however, is an explanation of why the US has fallen behind its competitors in the Arctic. This article fills in that gap by attempting to explain why the US is behaving as it does. It then argues that paradoxically falling behind in this regional competition may actually improve America's overall security and international influence when compared to Russia and China.

## Geography

The US is relatively fortunate in its geography. It has large coastlines with natural harbors on both the Pacific and Atlantic Oceans. Its rivers largely flow southward to southern ports. It also shares borders with Mexico and Canada, two countries that do not threaten the US in a conventional sense. This geography serves to protect the US from foreign invasion and allows it to readily deploy military forces to foreign locales, without use of the Arctic.

With the advent of intercontinental missiles and strategic bombers, the Arctic became more important to the US militarily during the Cold War. This pushed the US to erect now largely defunct early warning stations across northern Alaska, Greenland, and Canada. More recently, it established incipient missile defense systems in the Arctic to deal with increased threats emanating from Russia, China, and North Korea and improved its ability to monitor the region. However, these systems have never been designed to control the Arctic, but instead to protect America, and its NATO allies, from foreign military threats coming from, or through, the Arctic. This is an important distinction.

From Center for International Maritime Security (CIMSEC), April 13 © 2023. Reprinted with permission. All rights reserved.

Russia does not share America's fortunate geographic position. Instead, its geographic positioning and acrimonious international relationships have pushed it to "conquer the Arctic." It has few "warm-water" ports and shares large land borders with many adversarial states. Russia's only ports that are free from year-round ice are located in Sevastopol (Crimea), Tartus (Syria), and in the Baltic and Barents Seas. Significantly, Russia has recently fought to maintain control over Sevastopol and Tartus, but still faces possible blockades by adversarial forces in the Black Sea, Mediterranean Sea, and Baltic Sea. Ukraine's attempt to join NATO, Finland's recent accession to the alliance and Sweden's standing bid to join, along with the West's attempts to overthrow Russia's surrogate in Syria, Bashar Assad, have heightened Russia's longstanding fear in this regard.

As a result, since the disastrous Russo-Japanese War of 1905, and especially during WW I and WW II,* and the Cold War, Russia has militarized the Arctic. This is something that it has taken up with renewed vigor under Vladimir Putin's regime. Russia's militarization of the Arctic has especially occurred in two spots. The first one is the ice-free Barents Sea, which Russia has relied on to access the world's oceans so that it can better protect its territory and international interests from foreign threats, and the second one is under the Arctic ice cap where its nuclear submarines have an icy bastion that protects them from NATO forces.

## Economics

The US largely has a free-market economy with strong interest groups that challenge its willingness to expand its commercial footprint in the Arctic. This has overwhelmingly kept it from attempting to control the Arctic like Russia has done and China is increasingly attempting to do. It is important to look at the times when American commercial interests have focused on the Arctic to understand America's overall lack of interest in this region. The three times the US has been economically drawn to the Arctic were to exploit temporarily scarce resources. This occurred with whale oil and seal skins during the 18th and 19th century, gold at the end of the 19th century, and oil during the mid-twentieth century. These intense periods of economic interest in the Arctic resulted in America's purchase of Alaska from Russia in 1867 and the development of Alaska in the decades afterwards. Notably, however, it is expensive and difficult to operate in the Arctic. As Canadian Arctic expert, Michael Byers highlights, even as the Arctic ice slowly melts, the region remains in complete darkness for half of the year and melting ice is dangerously unpredictable. The Arctic is also austere and quite far from the largest population centers of the world. As such, the intermittent economic demands for the region's natural resources have relatively quickly resulted in substitutes being found for these goods in less austere places.

Subsequently, the only portions of Alaska that are significantly developed are in the sub-Arctic portion of the state, with the exception of the oil fields of Prudhoe Bay—which also appear to be winding down with the advent of fracking and renewable energy. Increasing environmental concerns (most of Alaska is situated in nationally owned wilderness preserves) and native groups' claims prohibitively increase the price of resource extraction from most of Arctic Alaska even further. Many Americans believe the region should be left to nature and to indigenous groups. The US also does not have a great need to develop the sea routes in the Arctic to improve its international trade. It has a transnational road and railway system and easy access to maritime trade routes which are connected through the recently enlarged Suez Canal. These circumstances mean that the US has very little motivation to establish sea routes through the largely uninhabited, relatively shallow, and dangerously unpredictable Arctic Ocean. Finally, Russia's aggression over the last two decades, and increasing pressure from environmentally-based NGOs, have pushed American-based companies even further away from Russia's Arctic.

All told, since the US has only marginal economic incentives to pursue the Arctic, it has not felt the need to develop harbors, settlements, transport infrastructure, or icebreakers to increase its footprint in the region. As such, it has relatively little capability to "conquer the region," but also relatively little to defend in the region.

This is not the case for Russia or China. Russia suffers from what Hill and Gaddy call the Siberian Curse. Its geography is not as economically favorable as America's, which has forced it to turn towards the Arctic to improve its economic circumstances. However, it has also traditionally operated a state-controlled economy that uses slave labor and nationally owned corporations to mask the economic, environmental and demographic costs of operating in the Arctic. Beginning with the czars, and accelerating under Russia's Soviet dictators, Russia forcibly sent millions of people to develop and "conquer the Arctic."

This legacy continues today as Putin pushes and subsidizes Russia's economic ministries and state-controlled corporations to extract more resources from the Arctic and to expand the infrastructure of the Northern Sea Route (with the numerous powerful icebreakers needed to navigate this waterway) to transport these resources to distant markets. Unlike American corporations, Russia's economic pursuits in the Arctic are not concerned with environmental or indigenous considerations either. Furthermore, Russia's extreme sacrifices in the Arctic have made developing and controlling it symbolic for its people and leadership. As such, Russia has much more to defend materially and ideationally in the Arctic than the US does. Even with these factors pushing Russia to conquer the Arctic, Russia's regional ambitions have been challenged by fiscal, demographic, and environmental hurdles. Most recently, the war in Ukraine has forced it to curtail its ambitious Arctic railway and icebreaker projects and to mobilize and sacrifice a significant proportion of its Arctic troops for combat in Ukraine. Additionally, many of its Arctic cities have rapidly de-populated, and the Arctic melt has paradoxically threatened its existing Arctic infrastructure.

Like Russia, China's companies are largely nationalized and it also does not have the environmental or indigenous concerns in the Arctic that the US does. It has spent the last two decades increasing its manufacturing sector and its international trade ties. This has increased its needs for natural resources and trade routes, resulting in its plans to establish a "Polar Silk Road," under its greater Belt and Road Initiative, in order to link the Arctic to China's greater network of international trading posts and manufacturing centers. As Russia has lost access to Western markets and technology over the last two decades, it has increasingly turned towards an eager China to help it build out its Arctic economic footprint. As such, China also has more economic interests to defend in the Arctic than the US does.

## What Does This Mean for the US?

The United States is not truly interested in competing for the Arctic. It has relatively less military, economic, or ideational interest in the region when compared to Russia or China. Its strategic plans for the region have become increasingly assertive in reaction to Russia's and China's efforts, but lack funding or prioritization. However, this lack of genuine interest carries some benefits for the US when considering the larger geopolitical context of the international system.

America's lack of interest in the region has paradoxically pushed the other Arctic states to increase their security ties with the US and to take on more security responsibilities for the region. Similar to World War II, when Iceland and Denmark invited the US to help protect their territory from foreign adversaries, Russia's aggression pushed Sweden and Finland to formally petition to join the US-dominated NATO. The inclusion of these states into the organization means that half of the Arctic will soon be administered by NATO member states.

Specifically, the Nordic states of Norway, Sweden and Finland have significant capabilities and economic stakes in the region that will make up for America's relative lack of willingness and ability to contain Russia's and China's ambitions in the region. These countries' capabilities will be further complemented by Denmark and Canada, and the other non-Arctic NATO states that have recently increased their defense spending to deal with Russian aggression. This collective defense in the Arctic will allow the US to better focus on domains like space, cyberspace, the Americas, and the Indo-Pacific, which are more important than the Arctic to America's most critical national interests.

> **Russia's militarization of the Arctic has occurred in two spots: the ice-free Barents Sea, which Russia has relied on to access the world's oceans; and under the Arctic ice cap where its nuclear submarines have an icy bastion that protects them from NATO forces.**

Economically speaking, the Arctic will likely remain a backwater for market-driven economies for the foreseeable future. The relatively high costs of extracting resources and transporting goods from the Arctic means the region is unlikely to become much more attractive for Western companies, even if the ice continues to retreat (which has slowed in recent years) and icebreakers improve, except in times when specific resources are in sharp demand or when there are long-term bottlenecks in other trade routes.

The resources that Russia and China extract from the Arctic will contribute to the overall global supply of these resources and decrease their overall price for American consumers. As such, Americans will gain many of the benefits of Russia's and China's efforts in the Arctic while Russia and China absorb the costs. In the case of scarce rare-earth minerals that have spiked in demand and are monopolized by China, it appears Sweden may fill this void for the US with its own Arctic resources, even as companies search for substitutes for these critical resources.

Overall, the US should not ignore the Arctic, and it should put to rest the notion that this region is a unique zone of peace in an otherwise quite turbulent world. That being said, Americans should also not deem that losing the "race for the Arctic" will critically threaten America's larger national interests. By not attempting to compete head-to-head with Russia or China to "conquer" the region, the US has incurred some advantages against these competitors.

As the US has been reminded again in Iraq and Afghanistan, and through its observation of Russia's disastrous invasion of Ukraine, conquering territory comes with significant costs that can weaken the material strength and ideational attractiveness of a country. This, in turn, weakens a country's ability to secure its most significant national interests. The US should continue to diplomatically, militarily, and economically challenge Russia's and China's actions in the Arctic on humanitarian and environmental grounds, but it also should identify that China's and Russia's actions in the Arctic come with high economic and soft power costs that may relatively benefit the US. Doing so will allow the US to increase its ability to collectively defend its interests in the Arctic with its allies and to prioritize its attention and resources on domains that are more important to it than the Arctic.

\*Interestingly, the US was responsible for a significant portion of Russia's militarization of the Arctic during World War II and went from supplying friendly Russian forces through the Arctic during WW I to fighting them in the Russian Arctic after the Bolshevik Revolution.

## Print Citations

**CMS**: Caldon, Josh. "Why the US Is Losing the Race for the Arctic and What to Do About It." In *The Reference Shelf: U.S. National Debate Topic 2025–2026: Exploration & Development in the Arctic,* edited by Micah L. Issitt, 180–184. Amenia, NY: Grey House Publishing, 2025.

**MLA**: Caldon, Josh. "Why the US Is Losing the Race for the Arctic and What to Do About It." *The Reference Shelf: U.S. National Debate Topic 2025–2026: Exploration & Development in the Arctic,* edited by Micah L. Issitt, Grey House Publishing, 2025, pp. 180–184.

**APA**: Caldon, J. (2025). Why the US is losing the race for the Arctic and what to do about it. M. L. Issitt (Ed.), *The reference shelf: U.S. national debate topic 2025–2026: Exploration & development in the Arctic* (pp. 180–184). Grey House Publishing. (Original work published 2023)

# Bibliography

"About." *International Arctic Science Committee (IASC)*, 2025, www.iasc.info/about.

"About the Saami Council." *Saamicouncil.net*, 2025, www.saamicouncil.net/en/the-saami-council.

"About Us." *Arctic Institute*, 2025, www.thearcticinstitute.org/about-us/.

"Ancient DNA Sheds Light on Arctic Hunger-Gatherer Migration to North American around 5,000 Years Ago." *Science Daily*, 5 June 2019, www.sciencedaily.com/releases/2019/06/190605133522.htm.

"The Arctic." *Arctic Centre, University of Lapland*, 2025, www.arcticcentre.org/EN.

"Arctic Countries." *Arctic Review*, 2025, arctic.review/international-affairs/arctic-countries/#:~:text=their%20Arctic%20policies.-,Russia,important%20stake%20in%20the%20region.

"Arctic Exploration Timeline." *American Polar Society*, 2024, americanpolar.org/arctic-exploration-timeline/.

Aton, Adam, and Lesley Clark. "Trump Declares War on State Climate Law." *Politico*, 9 Apr. 2025.

Buschman, Victoria Qutuuq. "Arctic Conservation in the Hands of Indigenous Peoples." *Wilson Quarterly*, Winter 2022.

"Conflict and Geopolitical Issues in the Arctic." *Discovering the Arctic*, 2025, discoveringthearctic.org.uk/arctic-people-resources/resources-from-the-edge/conflicts-geopolitical-issues/.

Dunaway, Finis. *Defending the Arctic Refuge: A Photographer, an Indigenous Nation, and a Fight for Environmental Justice*. U of North Carolina P, 2021.

Evans, Jen, and Andreas Østhagen. "Fisheries Disputes: The Real Potential for Arctic Conflict." *Arctic Institute*, 3 June 2021, www.thearcticinstitute.org/fisheries-disputes-real-potential-arctic-conflict/.

Gjerstad, Michael, and James Rogers. "Knowledge is Power: Greenland, Great Powers, and Lessons from the Second World War." *Arctic Institute*, 15 June 2021.

Griggs, Mary Beth. "The First People to Settle Across North America's Arctic Regions Were Isolated for 4,000 Years." *Smithsonian Magazine*, 28 Aug. 2014.

Heidt, Amanda. "Did Humans Cross the Bering Strait after the Land Bridge Disappeared?" *Live Science*, 20 Dec. 2023, www.livescience.com/archaeology/did-humans-cross-the-bering-strait-after-the-land-bridge-disappeared#:~:text=According%20to%20John%20Hoffecker%2C%0a,during%20one%20of%20these%20stretches.

Huebert, Rob. "Can the Arctic Council Survive the Trump Administration? Probably Not. Here's Why." *Arctic Today*, 3 Mar. 2025, www.arctictoday.com/can-the-arctic-council-survive-the-trump-administration-probably-not-heres-why/.

———. "A New Cold War in the Arctic?! The Old One Never Ended!" *Arctic Yearbook*, 2009, arcticyearbook.com/arctic-yearbook/2019/2019-commentaries/325-a-new-cold-war-in-the-arctic-the-old-one-never-ended.

"The Indigenous World 2024: Sápmi." *International Working Group for Indigenous Affairs (IWGIA)*, 2025.

Jóhannesson, Magnús. "Arctic Council: Structure, Work and Achievements." *Arctic Circle*, 12 Dec. 2022, www.arcticcircle.org/journal/arctic-council-structure-work-and-achievements.

"King Henry III's Polar Bear." *Historic UK*, 2025, www.historic-uk.com/CultureUK/Henry-III-Polar-Bear/.

Klein, Naomi. *This Changes Everything: Capitalism vs. The Climate*. Simon & Schuster, 2014.

Laineman, Matti, and Juha Nurminen. *A History of Arctic Exploration: Discovery, Adventure and Endurance at the Top of the World*. Bloomsbury USA, 2009.

"The Long, Long Battle for the Arctic National Wildlife Refuge." *National Resources Defense Council (NRDC)*, 15 Mar. 2024, www.nrdc.org/stories/long-long-battle-arctic-national-wildlife-refuge.

Luse, Jeff. "Markets Don't Want More Coal: Trump Is Propping Up the Industry Anyway." *Reason*, 10 Apr. 2025, reason.com/2025/04/10/markets-dont-want-more-coal-trump-is-propping-up-the-industry-anyway/.

Martinez, Chris, et al. "These Fossil Fuel Industry Tactics Are Fueling Democratic Backsliding." *Center for American Progress*, 5 Dec. 2023.

McInnis, Kathleen, et al. "Pulling Their Weight: The Data on NATO Responsibility Sharing." *Center for Strategic and International Studies (CSIS)*, 22 Feb. 2024, www.csis.org/analysis/pulling-their-weight-data-nato-responsibility-sharing.

Murkins, Sydney. "The Future Battlefield Is Melting: An Argument for Why the U.S. Must Adopt a More Proactive Arctic Strategy." *Arctic Institute*, 3 Dec. 2024, www.thearcticinstitute.org/future-battlefield-melting-argument-us-must-adopt-more-proactive-arctic-strategy/.

"Oil Spill Cleanup Workers More Likely to Have Asthma Symptoms." *National Institutes of Health (NIH)*, 17 Aug. 2022, www.nih.gov/news-events/news-releases/oil-spill-cleanup-workers-more-likely-have-asthma-symptoms.

"Protect the Arctic." *Greenpeace*, www.greenpeace.org/usa/protect-the-arctic/.

"Protecting the Arctic." *Ocean Conservancy*, 2025, oceanconservancy.org/protecting-the-arctic/.

"Qikiqtaaluk Region." *Government of Nunavut*, 2024, www.gov.nu.ca/sites/default/files/documents/2024-08/03_-_Info_about_Qikiqtaaluk_Region.pdf.

Reid, Robert Leonard. *Arctic Circle: Birth and Rebirth in the Land of the Caribou*. David Godine Publishers, 2010.

Rockwell, Keith. "Criticism of NATO Ignores Its Economic Benefit to the US." *Wilson Center*, 29 Mar. 2024, www.wilsoncenter.org/article/criticism-nato-ignores-its-economic-benefit-us.

Roosevelt, Theodore. "Conservation as a National Duty." *Voices of Democracy,* 13 May 1908, voicesofdemocracy.umd.edu/theodore-roosevelt-conservation-as-a-national-duty-speech-text/.

Rowe, Mark. "Arctic Nations Are Squaring Up to Exploit the Region's Rich Natural Resources." *Geographical*, 12 Aug. 2022, geographical.co.uk/geopolitics/the-world-is-gearing-up-to-mine-the-arctic.

Sharp, Gregor. "A Brief History of Lines in the Arctic." *Arctic Institute*, 20 Mar. 2018, www.thearcticinstitute.org/brief-history-lines-arctic/.

Starchak, Maxim. "Russia's Arctic Policy Poses a Growing Nuclear Threat." *Carnegie Endowment*, 1 Nov. 2024, carnegieendowment.org/russia-eurasia/politika/2024/10/russia-arctic-nuclear-threat?lang=en.

"The State of Arctic Food." *Arctic Economic Council (AEC)*, 2023, arcticeconomiccouncil.com/wp-content/uploads/2023/09/aec-arctic-food-report-2023.pdf.

"Trump Signs Order to Protect Big Oil from State Emissions Fines." *Bloomberg*, 9 Apr. 2025, www.bloomberg.com/news/articles/2025-04-09/trump-signs-order-to-protect-big-oil-from-state-emissions-fines.

Voosen, Paul. "Trump Seeks to End Climate Research at Premier U.S. Climate Agency." *Science*, 11 Apr. 2025, www.science.org/content/article/trump-seeks-end-climate-research-premier-u-s-climate-agency.

"Warmest Arctic Summer on Record Is Evidence of Accelerating Climate Change." *National Oceanic and Atmospheric Administration (NOAA)*, 12 Dec. 2023, www.noaa.gov/news-release/warmest-arctic-summer-on-record-is-evidence-of-accelerating-climate-change.

"What Were the Arctic Convoys?" *Arctic Convoy Museum*, 2025, arcticconvoymuseum.org/history-learning/what-were-the-arctic-convoys/.

Winston, Andrew, and Hunter Lovins. "Fossil Fuel Jobs Will Disappear, So Now What?" *MIT Sloan Management Review*, 13 May 2021, sloanreview.mit.edu/article/fossil-fuel-jobs-will-disappear-so-now-what/.

"Working Together in a Changing Canadian Arctic." *ArcticNet*, 2025, arcticnet.ca/.

"WWF Global Arctic Programme." *Arctic World Wildlife Fund (WWF)*, 2025, www.arcticwwf.org.

# Websites

### Arctic Council
www.arcticcouncil.org
The Arctic Council is the premier intergovernmental organization focused on the Arctic and involving members from all the nations with Arctic territory and Indigenous organizations and/or groups with territory in the Arctic. The Arctic Council was formed in the 1990s, and has supported studies on Indigenous rights and welfare, climate change, and oil and gas resources. Students and researchers can use the Arctic Council website to read about current issues, past studies, and analyses on Arctic concerns.

### The Arctic Institute (TAI)
www.thearcticinstitute.org
The Arctic Institute (TAI) is a think tank founded in 2011 and headquartered in Washington, D.C., which supports research and produces analysis on Arctic issues. Consistently ranked as among the most influential think tanks in the world, TAI supports and conducts research on Arctic energy and resources, migration and urbanization, military and commercial policy in the Arctic ocean, and economic research on Arctic fishing. Some of the research conducted or supported by the think tank is made available, for free, through its website for use by students, journalists, and researchers.

### International Arctic Science Committee (IASC)
www.iasc.info
The International Arctic Science Committee (IASC) is a nongovernmental organization (NGO) founded in 1990 in Iceland that has representatives from all eight Arctic nations: Canada, Finland, Iceland, Norway, Denmark, Sweden, the United States, and Russia. The organization also involves representatives from twenty-four countries all involved in funding and supporting science in the Arctic. The IASC helps to promote international scientific cooperation and produced studies on various Arctic issues, including climate change, resources, and environmental exploration.

### Inuit Circumpolar Council (ICC)
www.inuitcircumpolar.com
The Inuit Circumpolar Council (ICC) is a multinational NGO representing the interests of the Inuit and Yupik people of Alaska, Canada, Greenland, and parts of Russia. The Council was created in the late 1970s and serves as a network of resources for Inuit and Yupik people living in different parts of the world. The

ICC supports a variety of research programs dealing with Indigenous rights, welfare, economics, and the preservation of resources.

### Polar Institute
www.wilsoncenter.org/program/polar-institute

The Woodrow Wilson International Center for Scholars (WWICS), better known as the "Wilson Center," is a Washington, D.C.-based think tank established in 1968. It is now part of the Smithsonian Institution. The Wilson Center supports and conducts research on a variety of governmental and policy issues, including economics and trade. Among the research and policy initiatives of the organization is the Wilson Center's Polar Institute which conducts and funds research into issues involving Arctic territories and US interests. The Polar Institute makes some of their research available for students. The Wilson Center was one of the organizations targeted by Donald Trump and allies in their effort to reduce the effectiveness of publicly supported scholarship. While the effectiveness of the organization is therefore in question, much of the previous scholarship conducted by the Wilson Center remains available through a variety of channels.

### Saami Council
www.now.org

The Saami Council is an international NGO representing the interests of the Sámi people from Finland, Norway, Russia, and Sweden. Founded in 1956, the Saami Council is one of the largest Indigenous organizations in Europe involved in supporting research and promoting policy regarding the Sámi people and their welfare in the Arctic territories.

### University of the Arctic (UArctic)
Saamicouncil.net

The University of the Arctic is an international network of universities, colleges, and other organizations involved in or promoting Arctic research and scholarship. The program was started in 2001 by the Arctic Council and has nearly 200 members representing a variety of nations with some level of interest in the Arctic region. After the April 2022 invasion of the Ukraine, the organization temporarily suspended the membership of fifty-five organizations representing the Russian Federation. UArctic makes a variety of research papers and programs partially available through their online portal.

### World Wildlife Fund (WWF)
www.wwf.org

The World Wildlife Fund (WWF) is one of the world's largest and most active NGOs in the realm of wildlife protection and preservation. The WWF has an active program in the Arctic dealing with a range of conservation issues. WWF informational reports and studies deal with issues like overfishing, the rising temperatures of the Arctic Ocean, and the threat of extinction for familiar Arctic species like the polar bear and arctic fox.

# Index

2022 Russian invasion of the Ukraine, xi, 7

Agreement on Cooperation on Aeronautical and Maritime Search and Rescue in the Arctic (2011), 16
Agreement on Cooperation on Marine Oil Pollution Preparedness and Response in the Arctic (2013), 16
Agreement on Enhancing International Arctic Scientific Cooperation (2017), 16
Agreement to Prevent Unregulated High Seas Fisheries, 17
Akeeagok, P.J., 89, 91
Akman, Peter, 90
Alaska Critical Mineral Research Center, 93
Alaska Division of Geological & Geophysical Surveys, 95
Alaska Land Transfer Acceleration Act, 102
Alaska LNG Project, 100
Alaska National Interest Lands Conservation Act (ANILCA), 102
Alaska Native Claims Settlement Act of 1971, 102
Alaska Native Vietnam-era Veterans Land Allotment Program, 102
Alaska Statehood Act of 1958, 102
Aleut International Association (AIA), 65, 66
Allied Powers, 158
American Daybreak, 56
Arctic Athabaskan Council, 65
Arctic Circle Assembly, 21
Arctic Coast Guard Forum, 16, 27
Arctic Council, xi, 7, 16, 17, 19, 21, 23, 24, 27, 28, 37, 62, 65, 66, 67, 68, 69, 70, 76, 134, 135, 163, 169, 174, 175, 178
Arctic Development Bank, 28
Arctic Economic Council, 24, 28, 80
Arctic Environmental Protection Strategy (AEPS), 6, 21, 27, 65
*Arctic Institute, 39, 158, 160, 163*
Arctic Investment Protocol, 28
Arctic Mayor's Forum, 24, 28
Arctic Monitoring and Assessment Program (AMAP), 27, 134
Arctic National Wildlife Refuge (ANWR), 82, 83, 100, 103, 141, 165
Arctic Report Card, 119, 126, 129, 132
Arctic Sustainable Energy Policy, 25
Arndt, Kyle, 49

Barents Euro-Arctic Council, 28
Bauer, Rob, 21
Belt and Road Initiative, 13, 17, 167, 177, 183
Bering Land Bridge, 3, 4
Biden, Joe, 16, 19, 60, 68, 139, 141
Boassen, Jørgen, 55
BRICS+ engagement, 174
Brown, Dryden, 60
Bureau of Land Management, 101, 102, 103
Buschman, Victoria Qutuuq, 116
Byers, Michael, 181

Cameron, Michael, 52
Camp Century, 33, 34, 123
Canada-U.S. agreement, 139
*Center for American Progress, 118*
Central Powers, 158
Chemnitz, Aaja, 62
Chinese-Russian trade, 17

climate change, x, xi, xii, 3, 5, 7, 8, 11, 13, 15, 16, 18, 19, 21, 23, 24, 25, 26, 30, 31, 37, 38, 39, 40, 41, 43, 44, 49, 50, 53, 59, 71, 76, 81, 83, 107, 115, 116, 117, 118, 119, 120, 122, 123, 124, 131, 132, 134, 135, 138, 139, 140, 142, 145, 146, 153, 154, 160, 164, 165, 169, 176
*Climate Wars: What People Will be Killed in the 21st Century,* 176
coastal flooding, 126
Coastal Plain Oil and Gas Leasing Program, 100, 101
Cold War, x, 5, 6, 11, 15, 16, 20, 21, 30, 31, 34, 80, 109, 123, 124, 134, 158, 159, 163, 165, 166, 177, 180, 181
*Colliers,* 32
Collins, Mike, 60
Columbus, Christopher, 79
Compact of Free Association, 61
Conservation of Arctic Flora and Fauna (CAFF), 27, 67
Cook, Frederick, 5
Cooperation on Marine Oil Pollution, Preparation and Response, 27
Cotton, Tom, 59

Dans, Tom, 56
Deepwater Horizon oil spill, 82
Department of Defense (DoD), 138, 164
Department of Energy (DOE), 95
*Detroit Free Press,* 31

Eagle, Agnico, 90
Egede, Múte B., 10
Emergency Prevention, Preparedness and Response (EPPR), 27
Erik the Red, 58, 79, 122, 146
Euro-Arctic Council, 24, 28
European and Mediterranean Plant Protection Organization (EPPO), 149

European Union (EU), xi, 16, 17, 25, 40, 139
*Exxon Valdez disaster,* 82

Fernandez, Jose, 62
Fernández, Marta Asenjo, 71
Fitzhugh, Bill, 3
Francis, Ernie, 49, 50, 52
Franklin, John, 4
Frederiksen, Mette, 10, 55
Frobisher, Martin, 4, 79

Geophysical Institute, 93
Gjerstad, Michael, 158
global warming, x, 3, 18, 34, 67, 105, 106, 107, 128, 134
Gray, Alexander, 61
*Greenland in the World—Nothing About Us Without Us,* 10, 11
Greenland Self-Government Act, 10, 13
Greenland-Iceland-United Kingdom (GIUK) Gap, 11, 20
Greenland-Svalbard-Norway (GSN) gap, 110
Greenpeace, 38
gross domestic product (GDP), 160
Gwich'in Council International, 65, 67

Henson, Matthew, 5
*High North News,* 17, 89, 90
HMS *Erebus,* 4
HMS *Terror,* 4
Homestead Act, 60
Hudson's Bay Company, 79

India's Arctic strategy, 173
Intercontinental ballistic missiles (ICBMs), 158
International Arctic Science Committee (IASC), 38
International Code for Ships Operating in Polar Waters (Polar Code), 18

International Convention for the Prevention of Pollution from Ships (MARPOL), 18
International Convention for the Safety of Life at Sea (SOLAS), 18
International Maritime Organization, 18
International Monetary Fund, 97
International Plant Protection Convention (IPPC), 149
Inuarak, Enookie, 46
Inuit Circumpolar Conference (now Inuit Circumpolar Council), 41, 65, 66
Israel-Hamas war, 105

Jarlov, Rasmus, 60
Jay Treaty, 68
Jensen, Erik, 61
Joint Norwegian-Russian Fisheries Commission, 27

Lauder, Estée, 59
Lauder, Ronald, 59
liquified natural gas (LNG), 99
Lose, Lars Gert, 59
Lovins, Hunter, 117

Maddox, Marisol, 134
Martinez, Chris, 118
Mary River Mine, 46, 90
McCarty, Jessica, 134
Mineral Industry Research Laboratory, 93
Mittimatalik Hunters & Trappers Organization, 46, 47
Monroe, James, 58
Murkins, Sydney, 160
Murphy, Patrick, 49
mutually assured destruction (or MAD), 6

National Aeronautics and Space Administration (NASA), 119

National Oceanic and Atmospheric Administration (NOAA), 119
National Park Service, 102, 103
National Petroleum Reserve, 101, 102, 141, 165
*National Research Strategy 2022–2023*, 13
National Resources Defense Council (NRDC), 83
National Security Council, 61
NATO Military Committee, 21
NATO Vilnius Summit Communique, 21
Naval Station Norfolk, 123
Nielsen, Rasmus Leander, 57
noise pollution, 46, 47
nongovernmental organizations (NGOs), xii, 37, 84
Nordenskiöld, Adolf Erik, 5
North Atlantic Treaty Organization (NATO), 11, 15, 16, 20, 21, 22, 55, 57, 59, 69, 87, 107, 108, 110, 159, 160, 161, 163, 166, 168, 169, 177, 180, 181, 183
Northern Forum, The, 23, 24, 28
Northern Sea Route (NSR), 17, 25, 26, 105, 138, 177, 182
Northwest Passage (NWP), 5, 11, 12, 79, 105, 138, 139, 157, 164
nuclear weapons, x, 5, 6, 80, 123

Ottawa Declaration, 7, 23, 24, 27, 28

Paleo-Eskimos, 3
Pearson, Jean Hanmer, 31
Peary, Robert, 5
People's Republic of China (PRC), 10, 12, 13, 17, 18, 55, 57, 59, 60, 88, 92, 94, 108, 138, 146, 163, 168, 173, 174, 177, 178, 180, 181, 182, 183, 184
Permafrost Pathways, 49, 51, 53
Permanent Participants, 16, 65, 66, 68, 69

Peter Thiel's Founders Fund, 60
Philberth, Karl and Bernhard, 30
Pillai, Ranj, 89
Pituffik Space Base (formerly Thule Air Base), 11, 58, 87, 166
Polar Security Cutter program, 19
Polar Silk Road, 13, 108, 163, 167, 177, 183
Protection of the Arctic Marine Environment (PAME), 27
Prudhoe Bay, 5, 82, 165, 181
Pursuing Opportunities for Long-Term Arctic Resilience for Infrastructure and Society (POLARIS) project, 44
Putin, Vladimir, 106, 107, 108, 109, 136, 167, 181, 182

Quakenbush, Lori, 131

Ræbild, Anders, 147
rare earth elements (REEs), 12, 86, 92, 97, 138, 139, 150, 160, 173
Regional Environmental Change, 43
*Revolve*, 71
Rogers, James, 158
Royal Greenland Trading Company, 79
Roy-Léveillée, Pascale, 50
Russian Association of Indigenous Peoples of the North (RAIPON), 65
Russian Federation, 65
Russo-Ukrainian conflict, xi, 7, 11, 15, 16, 17, 19, 24, 65, 69, 70, 106, 108, 134, 136, 164, 165, 166, 168, 176, 181, 182, 184

Saami Council, 40, 65
Sáami people, xi, 39
Save the Arctic campaign, 38
Scandinavian nations, x, xii
sector theory, 6
Simon, Mary, 65, 70
Simpson, R.J., 90

Solana, Mike, 60
Spur and Siding Constructors Company, 32
Submerged Lands Act of 1953, 103
Suez Canal, 105, 106, 107, 164, 167, 168, 182
Supplementary Defense Cooperation Agreement, 177
Surabian, Andy, 63
Sustainable Development Working Group, 69
Sustainable Forestry Protocol, 27

*This Changes Everything: Capitalism vs. The Climate*, 81
Transboundary Waters Protection Act, 140
Transpolar Sea Route (TSR), 11, 12, 105, 107
Trump, Donald, xi, xii, 7, 10, 41, 55, 63, 80, 83, 89, 99, 117, 118, 119, 122, 124, 138, 159
Trump, Donald, Jr., 55, 63

U.S. Arctic Region Policy, 163, 165, 166
U.S. Arctic Research Commission, 56, 60
U.S. Arctic strategy, 13, 23, 109, 110, 163, 164, 165, 168, 169, 177
U.S. Embassy, 59
U.S. European Command (USEUCOM), 110
U.S. Geological Survey (USGS), 12, 17, 93, 95, 163
U.S. Indo-Pacific Command (USINDOPACOM), 110
U.S. national security, 10, 16, 23, 27, 29, 92, 99, 100, 105, 123, 140, 163, 169
U.S. National Strategy for the Arctic Region, 19, 163, 164, 165, 168
U.S. Northern Command (USNORTHCOM), 110

*Undark*, 30, 49, 50, 51, 52, 122, 134
Union of Soviet Socialist Republics (USSR), 163
United Nations Commission on the Limit of the Continental Shelf (CLCS), 18
United Nations Convention on the Law of the Sea (UNCLOS), 6, 18, 106
United Nations Declaration on the Rights of Indigenous Peoples (UNDRIP), 75
United Nations Educational, Scientific and Cultural Organization (UNESCO), 76

Vance, JD, 10
Veraverbeke, Sander, 135

*Wall Street Journal*, 56, 61

War of 1812, 157
Weather War, 20
Welzer, Harald, 176
Williams-Derry, Clark, 143
Willow Project, 141, 142
*Wilson Center*, 20, 65, 92, 116
Winston, Andrew, 117
Woodwell Climate Research Center, 49
World Jewish Congress, 59
World War I, x, 80, 158, 181, 184
World War II, x, 5, 6, 20, 31, 58, 59, 62, 80, 122, 158, 181, 183, 184
World Wildlife Federation, 26
World Wildlife Fund (WWF), 38, 143

Yukon Flats National Wildlife Refuge, 67
Yukon Gold Rush, 81

# Titles from Salem Press

Visit www.SalemPress.com for Product Information, Table of Contents, and Sample Pages.

## LITERATURE

### Critical Insights: Authors

Louisa May Alcott
Sherman Alexie
Dante Alighieri
Isabel Allende
Maya Angelou
Isaac Asimov
Margaret Atwood
Jane Austen
James Baldwin
Saul Bellow
Roberto Bolano
Ray Bradbury
The Brontë Sisters
Gwendolyn Brooks
Albert Camus
Raymond Carver
Willa Cather
Geoffrey Chaucer
John Cheever
Kate Chopin
Joseph Conrad
Charles Dickens
Emily Dickinson
Frederick Douglass
T. S. Eliot
George Eliot
Harlan Ellison
Ralph Waldo Emerson
Louise Erdrich
William Faulkner
F. Scott Fitzgerald
Gustave Flaubert
Horton Foote
Benjamin Franklin
Robert Frost
Neil Gaiman
Gabriel Garcia Marquez
Thomas Hardy
Nathaniel Hawthorne
Robert A. Heinlein
Lillian Hellman
Ernest Hemingway
Langston Hughes
Zora Neale Hurston
Henry James
Thomas Jefferson
James Joyce
Jamaica Kincaid
Stephen King
Martin Luther King, Jr.
Barbara Kingsolver
Abraham Lincoln
C.S. Lewis
Mario Vargas Llosa
Jack London
James McBride
Cormac McCarthy
Herman Melville
Arthur Miller
Toni Morrison
Alice Munro
Tim O'Brien
Flannery O'Connor
Eugene O'Neill
George Orwell
Sylvia Plath
Edgar Allan Poe
Philip Roth
Salman Rushdie
J.D. Salinger
Mary Shelley
John Steinbeck
Amy Tan
Leo Tolstoy
Mark Twain
John Updike
Kurt Vonnegut
Alice Walker
David Foster Wallace
H. G. Wells
Edith Wharton
Walt Whitman
Oscar Wilde
Tennessee Williams
Virginia Woolf
Richard Wright
Malcolm X

### Critical Insights: Works

Absalom, Absalom!
Adventures of Huckleberry Finn
The Adventures of Tom Sawyer
Aeneid
All Quiet on the Western Front
All the Pretty Horses
Animal Farm
Anna Karenina
As You Like It
The Awakening
The Bell Jar
Beloved
Billy Budd, Sailor
The Bluest Eye
The Book Thief
Brave New World
The Canterbury Tales
Catch-22
The Catcher in the Rye
The Color Purple
Crime and Punishment
The Crucible
Death of a Salesman
The Diary of a Young Girl
Dracula
Fahrenheit 451
A Farewell to Arms
Frankenstein; or, The Modern Prometheus
The Grapes of Wrath
Great Expectations
The Great Gatsby
Hamlet
The Handmaid's Tale
Harry Potter Series
Heart of Darkness
The Hobbit
The House on Mango Street
How the Garcia Girls Lost Their Accents
The Hunger Games Trilogy
I Know Why the Caged Bird Sings
In Cold Blood
The Inferno
Invisible Man
Jane Eyre
The Joy Luck Club
Julius Caesar
King Lear
The Kite Runner
Life of Pi
Little Women
Lolita
Lord of the Flies
The Lord of the Rings
Macbeth
The Merchant of Venice
The Metamorphosis
Midnight's Children
A Midsummer Night's Dream
Moby-Dick
Mrs. Dalloway
Native Son
Nineteen Eighty-Four
The Odyssey
Of Mice and Men
The Old Man and the Sea
On the Road
One Flew Over the Cuckoo's Nest
One Hundred Years of Solitude
Othello
The Outsiders
Paradise Lost
The Pearl
The Plague
The Poetry of Baudelaire
The Poetry of Edgar Allan Poe
A Portrait of the Artist as a Young Man
Pride and Prejudice
A Raisin in the Sun
The Red Badge of Courage
Romeo and Juliet
The Scarlet Letter
Sense and Sensibility
Short Fiction of Flannery O'Connor
Slaughterhouse-Five
The Sound and the Fury
A Streetcar Named Desire
The Sun Also Rises
A Tale of Two Cities
The Tales of Edgar Allan Poe
Their Eyes Were Watching God
Things Fall Apart
To Kill a Mockingbird
Twelfth Night, or What You Will
Twelve Years a Slave
War and Peace
The Woman Warrior
Wuthering Heights

Grey House Publishing | Salem Press | H.W. Wilson | 4919 Route 22, PO Box 56, Amenia NY 12501-0056

**SALEM PRESS**

# Titles from Salem Press

**SALEM PRESS**

Visit www.SalemPress.com for Product Information, Table of Contents, and Sample Pages.

## Critical Insights: Themes
The American Comic Book
American Creative Non-Fiction
The American Dream
American Multicultural Identity
American Road Literature
American Short Story
American Sports Fiction
The American Thriller
American Writers in Exile
Censored & Banned Literature
Civil Rights Literature, Past & Present
Coming of Age
Conspiracies
Contemporary Canadian Fiction
Contemporary Immigrant Short Fiction
Contemporary Latin American Fiction
Contemporary Speculative Fiction
Crime and Detective Fiction
Crisis of Faith
Cultural Encounters
Dystopia
Family
The Fantastic
Feminism Flash Fiction
Friendship
Gender, Sex and Sexuality
Going Into the Woods
Good & Evil
The Graphic Novel
Greed
Harlem Renaissance
The Hero's Quest
Historical Fiction
Holocaust Literature
The Immigrant Experience
Inequality
LGBTQ Literature
Literature in Times of Crisis
Literature of Protest
Love
Magical Realism
Midwestern Literature
Modern Japanese Literature
Nature & the Environment
Paranoia, Fear & Alienation
Patriotism
Political Fiction
Postcolonial Literature
Power & Corruption
Pulp Fiction of the '20s and '30s
Rebellion
Russia's Golden Age
Satire
The Slave Narrative
Social Justice and American Literature
Southern Gothic Literature
Southwestern Literature
The Supernatural
Survival
Technology & Humanity

Truth & Lies
Violence in Literature
Virginia Woolf & 20th Century Women Writers
War

## Critical Insights: Film
Bonnie & Clyde
Casablanca
Alfred Hitchcock
Stanley Kubrick

## Critical Approaches to Literature
Critical Approaches to Literature: Feminist
Critical Approaches to Literature: Moral
Critical Approaches to Literature: Multicultural
Critical Approaches to Literature: Psychological

## Literary Classics
Recommended Reading: 600 Classics Reviewed

## Novels into Film
Novels into Film: Adaptations & Interpretation
Novels into Film: Adaptations & Interpretation, Volume 2

## Critical Surveys of Literature
Critical Survey of American Literature
Critical Survey of Drama
Critical Survey of Long Fiction
Critical Survey of Mystery and Detective Fiction
Critical Survey of Poetry
Critical Survey of Poetry: Contemporary Poets
Critical Survey of Science Fiction & Fantasy Literature
Critical Survey of Shakespeare's Film Adaptations
Critical Survey of Shakespeare's Plays
Critical Survey of Shakespeare's Sonnets
Critical Survey of Short Fiction
Critical Survey of World Literature
Critical Survey of Young Adult Literature

## Critical Surveys of Graphic Novels
Heroes & Superheroes
History, Theme, and Technique
Independents & Underground Classics
Manga

## Critical Surveys of Mythology & Folklore
Creation Myths
Deadly Battles & Warring Enemies
Gods & Goddesses
Heroes and Heroines
Legendary Creatures
Love, Sexuality, and Desire
World Mythology

## Cyclopedia of Literary Characters & Places
Cyclopedia of Literary Characters
Cyclopedia of Literary Places

---

Grey House Publishing | *Salem Press* | H.W. Wilson | 4919 Route 22, PO Box 56, Amenia NY 12501-0056

# Titles from Salem Press

Visit www.SalemPress.com for Product Information, Table of Contents, and Sample Pages.

## Introduction to Literary Context
American Poetry of the 20th Century
American Post-Modernist Novels
American Short Fiction
English Literature
Plays
World Literature

## Magill's Literary Annual
Magill's Literary Annual, Annual Editions 1977-2024

## Masterplots
Masterplots, Fourth Edition
Masterplots, 2010-2018 Supplement

## Notable Writers
Notable African American Writers
Notable American Women Writers
Notable Horror Fiction Writers
Notable Mystery & Detective Fiction Writers
Notable Writers of the American West & the Native American Experience
Notable Writers of LGBTQ+ Literature

# HISTORY
## The Decades
The 1900s in America
The 1910s in America
The Twenties in America
The Thirties in America
The Forties in America
The Fifties in America
The Sixties in America
The Seventies in America
The Eighties in America
The Nineties in America
The 2000s in America
The 2010s in America

## Defining Documents in American History
Defining Documents: The 1900s
Defining Documents: The 1910s
Defining Documents: The 1920s
Defining Documents: The 1930s
Defining Documents: The 1950s
Defining Documents: The 1960s
Defining Documents: The 1970s
Defining Documents: The 1980s
Defining Documents: American Citizenship
Defining Documents: The American Economy
Defining Documents: The American Revolution
Defining Documents: The American West
Defining Documents: Business Ethics
Defining Documents: Capital Punishment
Defining Documents: Censorship
Defining Documents: Civil Rights
Defining Documents: Civil War
Defining Documents: Conservatism
Defining Documents: The Constitution
Defining Documents: The Cold War
Defining Documents: Dissent & Protest
Defining Documents: Domestic Terrorism & Extremism
Defining Documents: Drug Policy
Defining Documents: The Emergence of Modern America
Defining Documents: Environment & Conservation
Defining Documents: Espionage & Intrigue
Defining Documents: Exploration and Colonial America
Defining Documents: The First Amendment
Defining Documents: The Free Press
Defining Documents: The Great Depression
Defining Documents: The Great Migration
Defining Documents: The Gun Debate
Defining Documents: Immigration & Immigrant Communities
Defining Documents: The Legacy of 9/11
Defining Documents: LGBTQ+
Defining Documents: Liberalism
Defining Documents: Manifest Destiny and the New Nation
Defining Documents: Native Americans
Defining Documents: Political Campaigns, Candidates & Discourse
Defining Documents: Postwar 1940s
Defining Documents: Prison Reform
Defining Documents: The Salem Witch Trials
Defining Documents: Secrets, Leaks & Scandals
Defining Documents: Slavery
Defining Documents: Supreme Court Decisions
Defining Documents: Reconstruction Era
Defining Documents: The Vietnam War
Defining Documents: The Underground Railroad
Defining Documents: U.S. Involvement in the Middle East
Defining Documents: Voters' Rights
Defining Documents: Watergate
Defining Documents: Workers' Rights
Defining Documents: World War I
Defining Documents: World War II

## Defining Documents in World History
Defining Documents: The 17th Century
Defining Documents: The 18th Century
Defining Documents: The 19th Century
Defining Documents: The 20th Century (1900-1950)
Defining Documents: The Ancient World
Defining Documents: Asia
Defining Documents: Genocide & the Holocaust
Defining Documents: Human Rights
Defining Documents: The Middle Ages
Defining Documents: The Middle East
Defining Documents: Nationalism & Populism
Defining Documents: The Nuclear Age
Defining Documents: Pandemics, Plagues & Public Health
Defining Documents: Religious Freedom & Religious Persecution
Defining Documents: Renaissance & Early Modern Era
Defining Documents: Revolutions
Defining Documents: The Rise & Fall of the Soviet Union
Defining Documents: Treason
Defining Documents: Women's Rights

Grey House Publishing | Salem Press | H.W. Wilson | 4919 Route 22, PO Box 56, Amenia NY 12501-0056

# Titles from Salem Press

Visit www.SalemPress.com for Product Information, Table of Contents, and Sample Pages.

## Great Events from History
Great Events from History: American History, Exploration to the Colonial Era, 1492-1775
Great Events from History: American History, Forging a New Nation, 1775-1850
Great Events from History: American History, War, Peace & Growth, 1850-1918
Great Events from History: The Ancient World
Great Events from History: The Middle Ages
Great Events from History: The Renaissance & Early Modern Era
Great Events from History: The 17th Century
Great Events from History: The 18th Century
Great Events from History: The 19th Century
Great Events from History: The 20th Century, 1901-1940
Great Events from History: The 20th Century, 1941-1970
Great Events from History: The 20th Century, 1971-2000
Great Events from History: Modern Scandals
Great Events from History: African American History
Great Events from History: The 21st Century, 2000-2016
Great Events from History: LGBTQ Events
Great Events from History: Human Rights
Great Events from History: Women's History

## Great Lives from History
Great Athletes
Great Athletes of the Twenty-First Century
Great Lives from History: The 17th Century
Great Lives from History: The 18th Century
Great Lives from History: The 19th Century
Great Lives from History: The 20th Century
Great Lives from History: The 21st Century, 2000-2017
Great Lives from History: African Americans
Great Lives from History: The Ancient World
Great Lives from History: American Heroes
Great Lives from History: American Women
Great Lives from History: Asian and Pacific Islander Americans
Great Lives from History: Autocrats & Dictators
Great Lives from History: The Incredibly Wealthy
Great Lives from History: Inventors & Inventions
Great Lives from History: Jewish Americans
Great Lives from History: Latinos
Great Lives from History: LGBTQ+
Great Lives from History: The Middle Ages
Great Lives from History: The Renaissance & Early Modern Era
Great Lives from History: Scientists and Science

## History & Government
American First Ladies
American Presidents
The 50 States
The Ancient World: Extraordinary People in Extraordinary Societies
The Bill of Rights
The Criminal Justice System
U.S. Court Cases
The U.S. Supreme Court

## SOCIAL SCIENCES
Civil Rights Movements: Past & Present
Countries, Peoples and Cultures
Countries: Their Wars & Conflicts: A World Survey
Education Today: Issues, Policies & Practices
Encyclopedia of American Immigration
Ethics: Questions & Morality of Human Actions
Issues in U.S. Immigration
Principles of Sociology: Group Relationships & Behavior
Principles of Sociology: Personal Relationships & Behavior
Principles of Sociology: Societal Issues & Behavior
Racial & Ethnic Relations in America
Weapons, Warfare & Military Technology
World Geography

## HEALTH
Addictions, Substance Abuse & Alcoholism
Adolescent Health & Wellness
Aging
Cancer
Community & Family Health Issues
Integrative, Alternative & Complementary Medicine
Genetics and Inherited Conditions
Infectious Diseases and Conditions
Magill's Medical Guide
Men's Health
Nutrition
Parenting: Styles & Strategies
Psychology & Behavioral Health
Social Media & Your Mental Health
Teens: Growing Up, Skills & Strategies
Women's Health

## Principles of Health
Principles of Health: Allergies & Immune Disorders
Principles of Health: Anxiety & Stress
Principles of Health: Depression
Principles of Health: Diabetes
Principles of Health: Hypertension
Principles of Health: Nursing
Principles of Health: Obesity
Principles of Health: Occupational Therapy & Physical Therapy
Principles of Health: Pain Management
Principles of Health: Prescription Drug Abuse
Principles of Health: Whole Body Wellness

## BUSINESS
Principles of Business: Accounting
Principles of Business: Economics
Principles of Business: Entrepreneurship
Principles of Business: Finance
Principles of Business: Globalization
Principles of Business: Leadership
Principles of Business: Management
Principles of Business: Marketing

*Grey House Publishing | Salem Press | H.W. Wilson | 4919 Route 22, PO Box 56, Amenia NY 12501-0056*

# SALEM PRESS

# Titles from Salem Press

# SALEM PRESS

Visit www.SalemPress.com for Product Information, Table of Contents, and Sample Pages.

## SCIENCE
Ancient Creatures
Applied Science
Applied Science: Engineering & Mathematics
Applied Science: Science & Medicine
Applied Science: Technology
Biomes and Ecosystems
Digital Literacy: Skills & Strategies
Earth Science: Earth Materials and Resources
Earth Science: Earth's Surface and History
Earth Science: Earth's Weather, Water and Atmosphere
Earth Science: Physics and Chemistry of the Earth
Encyclopedia of Climate Change
Encyclopedia of Energy
Encyclopedia of Environmental Issues
Encyclopedia of Global Resources
Encyclopedia of Mathematics and Society
Environmental Sustainability: Skills & Strategies
Forensic Science
Notable Natural Disasters
The Solar System
USA in Space

## *Principles of Science*
Principles of Aeronautics
Principles of Anatomy
Principles of Archaeology
Principles of Architecture
Principles of Astronomy
Principles of Behavioral Science
Principles of Biology
Principles of Biotechnology
Principles of Botany
Principles of Chemistry
Principles of Climatology
Principles of Computer-aided Design
Principles of Computer Science
Principles of Cybersecurity
Principles of Digital Arts & Multimedia
Principles of Ecology
Principles of Energy
Principles of Environmental Engineering
Principles of Fire Science
Principles of Food Science
Principles of Forestry & Conservation
Principles of Geology
Principles of Graphic Design & Typography
Principles of Information Technology
Principles of Marine Science
Principles of Mass Communication
Principles of Mathematics
Principles of Mechanics
Principles of Microbiology
Principles of Modern Agriculture
Principles of Pharmacology
Principles of Physical Science
Principles of Physics
Principles of Probability & Statistics
Principles of Programming & Coding
Principles of Robotics & Artificial Intelligence
Principles of Scientific Research
Principles of Sports Medicine & Exercise Science
Principles of Sustainability

Principles of Zoology

## CAREERS
Careers: Paths to Entrepreneurship
Careers in Archaeology & Museum Services
Careers in Artificial Intelligence
Careers in the Arts: Fine, Performing & Visual
Careers in the Automotive Industry
Careers in Biology
Careers in Biotechnology
Careers in Building Construction
Careers in Business
Careers in Chemistry
Careers in Communications & Media
Careers in Criminal Justice
Careers in Culinary Arts
Careers in Cybersecurity
Careers in Earth Science
Careers in Education & Training
Careers in Engineering
Careers in Environment & Conservation
Careers in Financial Services
Careers in Fish & Wildlife
Careers in Forensic Science
Careers in Gaming
Careers in Green Energy
Careers in Healthcare
Careers in Heavy Equipment Operation, Maintenance & Repair
Careers in Hospitality & Tourism
Careers in Human Services
Careers in Illustration & Animation
Careers in Information Technology
Careers in Intelligence & National Security
Careers in Law, Criminal Justice & Emergency Services
Careers in Mass Communication
Careers in the Music Industry
Careers in Manufacturing & Production
Careers in Medical Technology
Careers in Nursing
Careers in Physics
Careers in Protective Services
Careers in Psychology & Behavioral Health
Careers in Public Administration
Careers in Sales, Insurance & Real Estate
Careers in Science & Engineering
Careers in Social Media
Careers in Sports & Fitness
Careers in Sports Medicine & Training
Careers in Technical Services & Equipment Repair
Careers in Transportation
Careers in Travel & Adventure
Careers in Writing & Editing
Careers Outdoors
Careers Overseas
Careers Working with Infants & Children
Careers Working with Animals

Grey House Publishing | *Salem Press* | H.W. Wilson | 4919 Route 22, PO Box 56, Amenia NY 12501-0056

# Titles from H.W. Wilson

Visit www.HWWilsonInPrint.com for Product Information, Table of Contents, and Sample Pages.

## The Reference Shelf
Affordable Housing
Aging in America
Alternative Facts, Post-Truth and the Information War
The American Dream
Artificial Intelligence
Book Bans & Censorship
The Business of Food
Campaign Trends & Election Law
College Sports
Democracy Evolving
The Digital Age
Embracing New Paradigms in Education
Food Insecurity & Hunger in the United States
Future of U.S. Economic Relations: Mexico, Cuba, & Venezuela
Gene Editing & Genetic Engineering
Global Climate Change
Guns in America
Hacktivism
Hate Crimes
Health Conspiracies
Immigration & Border Control in the 21st Century
Income Inequality
Internet Abuses & Privacy Rights
Internet Law
Labor Unions
LGBTQ in the 21st Century
Marijuana Reform
Mental Health Awareness
Money in Politics
National Debate Topic 2020/2021: Criminal Justice Reform
National Debate Topic 2021/2022: Water Resources
National Debate Topic 2022/2023: Emerging Technologies & International Security
National Debate Topic 2023/2024: Economic Inequality
National Debate Topic 2024/2025: Intellectual Property Rights
National Debate Topic 2025/2026: The Arctic
New Developments in Artificial Intelligence
New Frontiers in Space
Policing in 2020
Pollution
Prescription Drug Abuse
Propaganda and Misinformation
Racial Tension in a Postracial Age
Reality Television
Renewable Energy
Representative American Speeches, Annual Editions
Reproductive Rights
Rethinking Work
Revisiting Gender
Russia & Ukraine
The South China Sea Conflict
Space Exploration
Sports in America
The Supreme Court
The Transformation of American Cities
The Two Koreas
UFOs
Vaccinations
Voting Rights
Whistleblowers

## Core Collections
Children's Core Collection
Fiction Core Collection
Graphic Novels Core Collection
Middle & Junior High School Core
Public Library Core Collection: Nonfiction
Senior High Core Collection
Young Adult Fiction Core Collection

## Current Biography
Current Biography Cumulative Index 1946-2025
Current Biography Magazine
Current Biography Yearbook

## Readers' Guide to Periodical Literature
Abridged Readers' Guide to Periodical Literature
Readers' Guide to Periodical Literature

## Indexes
Index to Legal Periodicals & Books
Short Story Index

## Sears List
Sears List of Subject Headings
Sears List of Subject Headings, Online Database

## History
American Game Changers: Invention, Innovation & Transformation
American Reformers
Speeches of the American Presidents

## Facts About Series
Facts About the 20th Century
Facts About American Immigration
Facts About China
Facts About the Presidents
Facts About the World's Languages

## Nobel Prize Winners
Nobel Prize Winners: 1901-1986
Nobel Prize Winners: 1987-1991
Nobel Prize Winners: 1992-1996
Nobel Prize Winners: 1997-2001
Nobel Prize Winners: 2002-2018

## Famous First Facts
Famous First Facts
Famous First Facts About American Politics
Famous First Facts About Sports
Famous First Facts About the Environment
Famous First Facts: International Edition

## American Book of Days
The American Book of Days
The International Book of Days

*Grey House Publishing | Salem Press | H.W. Wilson* | 4919 Route 22, PO Box 56, Amenia NY 12501-0056